AF799787

Elements of the Swiss Market for Electricity

Contributions to Economics

Peter R. Haiss
Cultural Influences on Strategic Planning
1990. ISBN 3-7908-0481-9

Manfred Kremer/Marion Weber (Eds.)
Transforming Economic Systems: The Case of Poland
1992. ISBN 3-7908-0585-8

Marcel F. van Marion
Liberal Trade and Japan
1993. ISBN 3-7908-0699-4

Hans Schneeweiß/
Klaus F. Zimmermann (Eds.)
Studies in Applied Econometrics
1993. ISBN 3-7908-0716-8

Gerhard Gehrig/
Wladyslaw Welfe (Eds.)
Economies in Transition
1993. ISBN 3-7908-0721-4

Christoph M. Schneider
Research and Development Management: From the Soviet Union to Russia
1994. ISBN 3-7908-0757-5

Bernhard Böhm/
Lionello F. Punzo (Eds.)
Economic Performance
1994. ISBN 3-7908-0811-3

Michael Reiter
The Dynamics of Business Cycles
1995. ISBN 3-7908-0823-7

Michael Carlberg
Sustainability and Optimality of Public Debt
1995. ISBN 3-7908-0834-2

Lars Olof Persson/Ulf Wiberg
Microregional Fragmentation
1995. ISBN 3-7908-0855-5

Andreas Foerster
Institutional Analysis of Development Administration
1995. ISBN 3-7908-0853-9

Ernesto Felli/Furio C. Rosati/
Giovanni Tria (Eds.)
The Service Sector: Productivity and Growth
1995. ISBN 3-7908-0875-X

Giuseppe Munda
Multicriteria Evaluation in Fuzzy Environment
1995. ISBN 3-7908-0892-X

Boris Maurer
R & D, Innovation and Industrial Structure
1996. ISBN 3-7908-0900-4

Giovanni Galizzi/Luciano Venturini (Eds.)
Economics of Innovation: The Case of Food Industry
1996. ISBN 3-7908-0911-X

David T. Johnson
Poverty, Inequality and Social Welfare in Australia
1996. ISBN 3-7908-0942-X

Rongxing Guo
Border-Regional Economics
1996. ISBN 3-7908-0943-8

Oliver Fratzscher
The Political Economy of Trade Integration
1996. ISBN 3-7908-0945-4

Ulrich Landwehr
Industrial Mobility and Public Policy
1996. ISBN 3-7908-0949-7

Arnold Picot/Ekkehard Schlicht (Eds.)
Firms, Markets, and Contracts
1996. ISBN 3-7908-0947-0

Karin Peschel (Ed.)
Regional Growth and Regional Policy Within the Framework of European Integration
1997. ISBN 3-7908-0957-8

Thorsten Wichmann
Agricultural Technical Progress and the Development of a Dual Economy
1997. ISBN 3-7908-0960-8

Ulrich Woitek
Business Cycles
1997. ISBN 3-7908-0997-7

Michael Carlberg
International Economic Growth
1997. ISBN 3-7908-0995-0

Massimo Filippini

Elements of the Swiss Market for Electricity

With 51 Figures
and 5 Tables

Physica-Verlag
A Springer-Verlag Company

Series Editors
Werner A. Müller
Peter Schuster

Author
PD Dr. Massimo Filippini
Institut für Empirische Wirtschaftsforschung (IEW)
University of Zürich
Blümlisalpstr. 10
CH-8006 Zürich, Switzerland

The publication of this study was partly financed by the Azienda Elettrica Ticinese, Bellinzona.

ISBN-13: 978-3-7908-0996-1

Cataloging-in-Publication Data applied for

Die Deutsche Bibliothek – CIP-Einheitsaufnahme
Filippini, Massimo: Elements of the Swiss market for electricity: with 5 tables / Massimo Filippini. – Heidelberg: Physica-Verl., 1997
(Contributions to economics)
ISBN-13: 978-3-7908-0996-1 e-ISBN-13: 978-3-642-47000-4
DOI: 10.1007/978-3-642-47000-4

Softcover Design: Erich Kirchner, Heidelberg

SPIN 10566040 88/2202-5 4 3 2 1 0 – Printed on acid-free paper

ACKNOWLEDGEMENTS

During the course of this research I have benefitted greatly from the advice and comments of many people. I wish to express my gratitude to Professor Peter Zweifel for providing me the opportunity to do this research while working at the Institute of Empirical Research in Economics at the University of Zürich. I owe special thanks to Professor Sybrand Schim Van Der Loeff, who taught me most about the practice of econometrics theory that underlies this research. Parts of the manuscript were written when I was visiting scholar at the Economics Department of Harvard University, Cambridge. I am grateful to Professors Ernst R. Berndt, Zvi Griliches, William W. Hogan, and Dale Jorgenson for their very helpful discussions and suggestions.

The acquisition of the data sets utilized in this research would not have been possible without the help and cooperation I received from the Swiss Electric Utilities Association, from Alfred Löhrer of the Swiss Energy Office and from research assistants Markus Meier and Paola Prioni. Their involvement is greatly appreciated.

I would also like to thank Professor Marc Hauser and research assistants Susanne Bonomo, Luca Crivelli, and Jörg Wild who read the manuscript with great care and offered constructive suggestions. In addition, I am also grateful to Janaki Blum, Tauni Sanchez and Tara Zend who have made the manuscript readable by scrutinizing the language.

Finally, I am gratefully to my parents, Eugenio and Annamaria, for their encouragement and to my wife and son, Angela and Simone, for their love and patience.

ACKNOWLEDGEMENTS

[illegible]

TABLE OF CONTENTS

LIST OF TABLES

LIST OF FIGURES

CHAPTER 1

INTRODUCTION

1.1 Economic issues to be analyzed

This research examines two elements of the Swiss market for electricity: the residential electricity demand by time-of-use and the cost structure of municipal electricity distribution utilities.

The empirical results of demand and cost elasticities allow the investigation of interesting economic and policy issues such as the desirability of a widespread introduction of time-of-use pricing for residential customers, the desirability of side-by-side competition in the distribution of electricity and, more generally, the economic effects of a reduction of the load factor and of mergers between electric distribution utilities on costs.

Desirability of time-of-use pricing

In the last decade there has been an intensifying debate in Switzerland about efficacy of electricity rate reforms in order to improve the efficiency of electricity use. This debate was initiated by two main events. First, there was an important growth of electricity consumption. Second, the Chernobyl accident in 1986 aroused widespread public concern about the problems associated with nuclear power and waste disposal. As a result, in 1991 the Swiss voted, in a referendum, a 10-year moratorium on the construction of new nuclear power plants. Moreover, plans to expand production of hydroelectric power (construction of new dams or expanding existing ones) have been stiffly opposed by environmental groups. These developments have consistently curtailed potential expansion of domestic electricity supply. As a result, Switzerland during the winter has to import electricity from foreign countries.

This evolution in the electricity market has increased interest in a widespread application of time-of-use pricing, which has long been proposed as an instrument to promote efficient utilization of existing electric plants and to slow the need for additional capacity. Moreover, time-of-use rates can be utilized to obtain an optimal domestic provision of electricity during the winter in a market with trade.[1]

Knowledge of the factors affecting electricity demand is, of course, very important for appraising time-of-use tariffs. The estimation of econometric models of electricity demand by time-of-use offers an interesting approach for studying the potential impact of alternative pricing policies. Moreover, knowledge of electricity demand allows analysis of residential consumer welfare implications if time-of-use tariffs are introduced or changed.

Unfortunately, the econometric literature on time-of-use demand consists of only a few studies.[2] Therefore, the need for new empirical evidence on this subject is high.

In brief, the purpose of the first part of this study is to provide an understanding of residential electricity demand by time-of-use with further empirical evidence of price and substitution elasticities for the Swiss residential sector.

The desirability of competition in the distribution of electricity

The success of deregulation in the transportation and communications sectors, as well as the privatization and competitive restructuring of the electricity supply sector in the United Kingdom, have

---

1 For an interesting discussion on the optimal domestic provision of electricity in markets with trade see Beck and Zweifel (1988, 1994).

2 For a overview of these studies see chapter 2 of this book and Hawdon (1992).

led to a number of recent proposals for the total or partial deregulation of the electric power industry in many countries. The possibility, therefore, of deregulating this industry is likely to become a very topical policy issue in Switzerland.[3]

The economic rationale for regulating the electric power industry is a widely held perception regarding the existence of scale economies in the production, transmission and distribution of electric power. This feature, of declining average costs in all three stages of electricity production, has been recognized by economists as one condition under which markets may fail to yield an efficient outcome. However, recent empirical research found evidence for exhaustion of economies of scale in the production of power, while they continue to exist in the transmission and distribution of electricity.[4] It seems essential that any such discussion of deregulation be prefaced by a clear understanding of the cost structure of the electric power industry. Moreover, the fact that electric distribution utilities are local monopolies that face cyclical demand and therefore are required to have a margin of spare capacity to satisfy demand during peak-periods, suggests an analysis of capacity utilization. In this respect, some economists claim that the spare capacity held by distributing utilities to meet peak-period demands is too high, reflecting over-capitalization.

The aim of the second part of this study is to provide an understanding of such problems with further empirical evidence on economies of scale, economies of utilization, cost elasticity with respect to load factor, and an analysis of over-capitalization in the Swiss electricity distribution industry.

The results of this study are relevant to such regulatory issues as whether the assumption of a natural monopoly in the distribution of electricity is valid, and whether continued regulation in terms of designation of service territory and, more generally, of the desirability of competition in the distribution of electric power in Switzerland is advisable.

---

3 See Hogan (1993) for an interesting discussion of a competitive electricity market.

4 See Christensen and Greene (1976) and Kamerschen and Thompson (1993) for empirical evidence of exhaustion of economies of scale in generation of power. See Weiss (1975), Henderson (1985), Roberts (1986), Nelson and Primeaux (1988), Callan (1991) and Salvanes and Tjφta (1994) for empirical evidence of the existence of scale economies in transmission and distribution of electricity.

Further, because time-of-use tariffs influence the load factor of electric utilities, empirical results regarding load factor elasticity allow an analysis of cost implications for electric distributors when introducing or changing, for instance, residential time-of-use tariffs. Moreover, the empirical results on the cost structure of Swiss electric utilities can be utilized to simulate the economic effects of mergers. The results of this simulation exercise, shortly illustrated in chapter 10, shows the possible cost savings that can be obtained by merging several electric distribution utilities. It is believed that this type of policy simulation exercise can help policy makers understand the economic impacts of merger policies.

1.2 Market for electricity: a short overview of demand, supply, regulation and tariff structures

In this section we briefly present some elements of the market for electricity in Switzerland. The main characteristics of demand, technical, market, regulation and tariff structures are highlighted.

Demand

In Switzerland the most important energy sources utilized by the industrial, residential and service sectors are oil, electricity and gas. In 1993 oil accounted for 62.9, electricity for 21, gas for 10.9, and coal, wood and other sources for 5.2 percent of total energy consumption.[5] Therefore, electricity is an important energy source in Switzerland.

Table 1.1 shows electricity consumption by sector in 1993. The residential sector is the second most important consumer of electricity and accounts for 30 percent of the total electricity consumption. Therefore, the pattern and development of residential electricity consumption can strongly influence total electricity consumption. Further, in the residential sector electricity is primarily used for space and water heating, and cooking purposes.

---

5 See Swiss Energy Office (1994).

Total electricity demand and peak electricity demand, which is the chief determinant of the need for new capacity, steadily increased during the last decades.[6] Since 1982 the total electricity consumption and peak electricity demand growth averaged 3 and 2.2 percent annually, respectively, with a total increase from 1983 to 1993 of approximately 25 and 23 percent, respectively. The growth of electricity consumption in the residential sector during the same period was higher than the growth of total consumption with an increase of approximately 36 percent.[7]

Table 1.1 Electricity consumption by sector in Switzerland (1993)

Residential	*Industry*	*Service*	*Agricultural*	*Transport*
30%	34.3%	25.4%	2%	8.3%

Source: Swiss Energy Office (1994)

Generally, electricity is a non-storable good with cyclic demand. This implies that electricity has to be produced when it is demanded. In Switzerland the electricity demand tends to be cyclic in two ways. First, due to the cold climate of Switzerland, winter electricity demand is higher than summer electricity demand. For instance, in 1993 the winter demand was approximately 23 percent higher than the summer demand. It was observed that during some of the last winters, the electricity demand was higher than the indigenous electricity supply, causing Switzerland to import electricity. Second, because most of the electricity consuming activities of the residential, industrial and service sectors occur during the day, daytime demand is higher than nighttime demand. Figure 1.1 shows a typical daily system load curve for Switzerland, which is obtained by plotting on a time axis the total electricity consumption at a given hour.

6 Peak demand is defined as the maximum level of electricity consumption at a given point in time and is measuredin Kilowatts (KW).

7 See Swiss Energy Office (1994).

In Switzerland the daytime demand is approximately 28 percent higher than the night-time demand. Therefore, during the time of peak demand the utilization of generating and distributing capacity is at a maximum, whereas during the period of off-peak demand part of this capacity is idle.

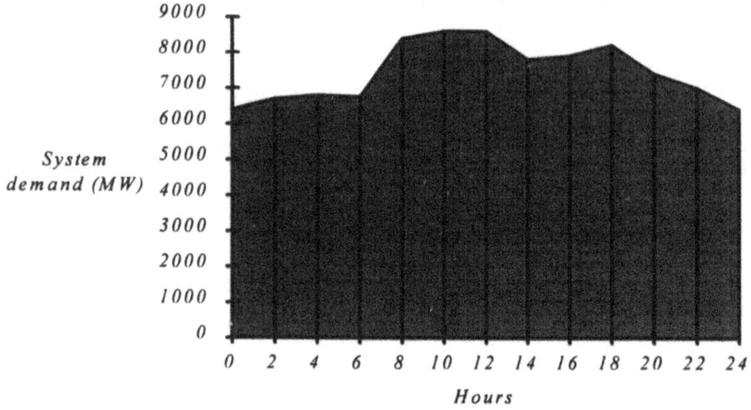

Figure 1.1 Typical daily system load curve for Switzerland

In sum, we can say that the electricity consumption pattern in Switzerland is characterized by steadily growth of total and peak consumption and by a cyclic demand that varies between the time of day and the seasons.

Technical structure

The production and distribution of electric power is complex, both physically and institutionally. Figure 1.2 presents a simple electric power system. This system consists of three functions: the generation of electricity, its transmission, and its distribution. Electricity is generated by three distinct processes: steam power, hydropower and nuclear power.

Generally, the production of electricity requires the presence of water sources. This implies that electric power has to be transported from the location of production to the location of final consumption. The transmission function is involved with the carriage of electric power from generating to distribution units. This transport is carried out over high voltage transmission lines which, in comparison to low voltage transmission lines, have little loss of power.

The distribution function is thus characterized by the delivery of electricity to the final users. Using transformers the electricity is stepped down to a lower voltage more appropriate for the final users. Therefore, the difference between the transmission and the distribution function is one of the voltage.

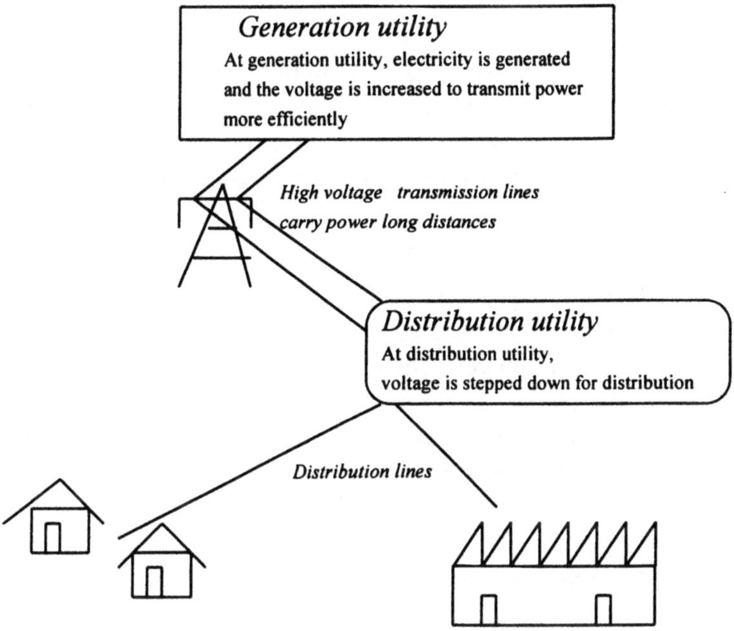

Figure 1.2 A simple electric system

Generally, we can distinguish three types of electric companies. The first is characterized by those that generate and transmit electricity. The second type is represented by electric companies which primarily distribute electricity. Finally, the third type of company performs all three electric power system functions and is vertically integrated.

Market structure

Switzerland is a modern federal state. The federal government is responsible for foreign policy, defense, pensions, post, telecommunications, railways, national highways and currency. What is not specifically in the hands of the federal government is, in principle, the responsibility of the cantons. This includes schooling, roads, hospitals, taxation systems, and electric utilities. Each canton comprises many municipalities, which not only implement such cantonal directions, as civil law and schooling, but also collect their own income taxes and chose which basic utilities to provide, such as water and electricity. Thus, because Switzerland is a federal state, the generation, transmission and distribution of electric power is carried out by private utilities or public utilities owned by the municipalities or the cantons. The organization of electric power production in Switzerland differs substantially from that of a centralized state like Italy or France where the generation, transmission and distribution of electricity is carried out by a big national public utility.

The Swiss electric power industry is composed of about 1,200 firms, public and private, that are engaged in the generation, transmission and/or distribution of electric power. There is great divergence in size and activity among these companies. In terms of numbers, utilities exclusively engaged in the distribution of electric power are dominant, with 74% of the total. The majority of these 900 or so companies are municipals, and provide power to their communities exclusively. The remaining 300 utilities operate within urban or regional areas. This group of firms is involved in generation, transmission and distribution, but generally the amount of generated power is small and is determined by the ability to exploit favorable hydroelectric power generation possibilities. The municipal and regional electric utilities purchase power mainly from 10 utilities which form the backbone of the industry. These larger vertically integrated companies provide most of the generated electricity and are also involved in the transmission and distribution of electricity to final consumers and municipal utilities. Moreover, these dominant companies own and control the transregional grid that is planned and used in close cooperation.

The size of these electric utilities vary from small municipal ones selling 100,000 kWh (Kilowatt-hours) to large utilities selling over 10,000,000 kWh. Approximately 300 utilities account for 95 percent of total electricity sales to end-users and approximately 900 utilities for the remaining 5 percent. Moreover, in Switzerland, the average number of persons served by an electric utility is approximately 6000, whereas in the United States it is about 100,000.

In term of ownership, approximately 30 percent of the total capital invested in the electric power industry is held by private investors, whereas 70 percent is owned by the cantons and the municipals. Table 1.2 provides a general overview of the industry structure described above.

In brief, the electric power industry in Switzerland has two main characteristics. First, a small number of large private electric utilities generate and distribute a high percentage of total sales of electricity to urban and municipal electric utilities and to end-users. Second, a large number of small public electric utilities owned by the cantons or municipals essentially distribute electric power to their communities.

Table 1.2 The electric power industry in Switzerland

	GTD utilities[a]	*GD utilities*[b]	*D utilities*[c]
Number	10	300	900
Output, measured by sales (in %)	50	45	5
Public utilities (in %)	70	90	95
Private utilities (in %)	30	10	5

Source: Mutzner (1989)

a. Utilities operating in the generation, transmission and distribution of electric power.
b. Utilities principally operating in the distribution of electric power but with a small part of their activities in the generation of electric power.
c. Utilities operating in the distribution of electric power.

Regulation structure

The electric power industry has always been regulated. The most important argument to support the economic regulation of this industry has been the pervasive belief that electric utilities are natural monopolies in their service areas. This belief led many governments of Western countries to regulate the activities of electric utilities.

In Switzerland, the federal government, the cantons and the municipalities exercise jurisdiction over the activities of the electric power industry. The federal government has authority to license and regulate the construction and operation of all nuclear power plants, the construction of transregional grids and the exploitation of water resources for hydroelectric power generation. Whereas, cantons and municipalities regulate the activities of regional private and public electricity utilities. The canton and municipal authorities regulate these utilities in different ways and only broad generalizations about their operations can be made. Some cantons regulate all utilities, while others limit jurisdiction to private utilities and leave regulation of local public or private utilities to local government. The regulatory laws on the canton and municipal levels govern entry, quality and condition of service including the obligation to serve all customers in the assigned service area and not to discriminate. Further, electricity price changes by private electric utilities have to be approved by the Federal Commission for the Supervision of Prices. Whereas electricity price changes by cantonally and municipally owned public electric utilities have to be approved by either the canton or the municipal public utility commission. In these cases, the intervention by federal price-supervisor commission is not always required.

The main electricity distribution policy option employed on canton and municipal level are: provision of electricity by a regulated public monopoly, provision of electricity by a privately owned but regulated monopoly franchise. In exchange for guaranteed service to all customers, public and private electric utilities receive an exclusive territorial franchise. Thus, these utilities operate as local monopolies in their legally defined service territories and are protected from competition from other utilities offering the same service in the same service area.

In practice the communes have the following two options when deciding how to provide their communities with power. The first option consists in the attribution of the exclusive territorial franchise to a private utility, while the second one involves the attribution to a local or cantonal public utility. This self-governing process in the organization of the distribution of electricity at the communal level can explain the high number of municipal electric utilities in Switzerland.

Tariff structure

In economics, theoretical principles underlying efficient allocation of goods suggest the implementation of marginal cost pricing. Therefore, electricity should be supplied up to the point at which the marginal cost is equal to the amount the consumer is willing to pay. Moreover, the fact that electric utilities face demands that vary according to hour and season introduces another dimension into the analysis of efficient pricing in the electricty sector. Because electricity is non-storable, utilities cannot meet peak demand with electricity produced in off-peak periods. Instead, electric utilities must always be able to satisfy demand during peak periods. This implies an over-capacity in off-peak periods and marginal cost that vary with changes in demand. The pattern of marginal cost of supplying electricity depends importantly on options available for generating power.[8]

Conventionally, an electric utility uses a variety of generating units characterized by different operating costs covering the supply of electricity during peak and off-peak periods. These utilities usually employ generating units with higher operating costs to meet peak demand because the peak period only lasts a few hours each day. Therefore, from a theoretical point of view, the most efficient pricing policy is to set price equal to marginal cost in each period. Moreover, fixed costs should be recovered by a fixed uniform customer service fee. Introduction of peak-load pricing would encourage shifting consumption to off-peak times and result in more efficient use of the capacity.

In periods of growing demand the investments in additional capacity are based on predictions of future peak demand. This implies, that principally only the increase of the peak demand is responsible for the cost of installing as well as operating any required capacity. Therefore, in order

8 For an optimal pricing policy in electricity markets with trade see Beck and Zweifel (1988, 1994).

to provide consumers with price signals that promote an efficient use of electricity, peak electricity prices should be based not only on historical but also on future cost. In this case consumers would be confronted with the full costs to society of using electricity during peak period.

In practice electric utility pricing decisions have to be reached under certain political, financial and administrative constraints.[9] Given the complexity of the issue, the traditional approach to electric utility management has relied on financial rather than economic criteria. As a result, the electricity tariffs designed by the Swiss electric utilities show homogeneity for industrial customers and substantial heterogeneity for residential customers.[10] Some utilities base the electricity pricing decision on the marginal cost principle, thus providing price signals for efficient energy use. Some others tend to view the rate design as more of an art than a science with the result of complex and inefficient tariffs.

The characteristics of Swiss electricity tariffs range from time invariant pricing such as block and two-part tariffs, to time-variant pricing such as time-of-day and two-part time-of-day tariffs. The latter have become the norm for industrial customers. Therefore, industrial customers are provided with price signals that reflect the time variant marginal cost influencing their use of electricity in an efficient way.

The residential electricity tariff designed by the Swiss electric utilities are numerous. Table 1.3 shows the most important types of residential electricity tariffs utilized by Swiss electric utilities.

One third of the Swiss electric utilities apply to the residential sector a two-part tariff, which consists of a monthly fixed fee plus a constant price per kWh of electricity consumed. The majority of electric utilities in Switzerland, however, use a combination of time-of-use tariffs and two-part tariffs. Two-part time-of-day tariff is a two-part tariff containing a fixed monthly charge and a constant price per kWh of electricity consumed that

---

9 See Anderson and Turvey (1977) for a presentation of practical applications of electricity tariffs.

10 The term electricity rate means the price charged per unit of electricity consumed. The term rate structure means the full set of prices charged for different units of electricity consumed by a customer. The term electricity tariff means both rate structure and all conditions that are associated with it.

varies according to time (day/night).[11] Therefore, this type of tariff reflects the pattern of variation in marginal cost and the presence of high fixed costs. Furthermore, we can distinguish electric utilities offering two-part time differentiated tariffs to all residential customers and electric utilities offering this type of tariff only to customers who use electricity extensively during the off-peak period, for example for electric space or water heating. Therefore, for these utilities we can observe customers on a two-part rate schedule and customers on a two-part time differentiated rate schedule.

Table 1.3 Residential electricity tariffs in Switzerland

Two-part tariffs	*Two-part Time-of-day tariffs*	*Block tariffs*	*Tariffs differentiated by type of use*
31 %	58 %	3 %	8 %

Source: Meier (1993)

Under block tariffs the electricity price changes discretely at the block boundaries. We distinguish increasing and decreasing block tariffs. With increasing block tariffs the marginal price increases as quantity consumed increases, whereas with decreasing block tariffs the marginal price decreases as quantity consumed increases. The utilities which apply tariffs differentiated by usage, charge a price per kWh of electricity consumed that varies according to the usage. For example, electricity for cooking usage has a different price than electricity for heating usage. It is notable that these latter two types of tariffs are no longer popular in Switzerland.

---

11 The time schedules for peak and off-peak electricity consumption applied by the Swiss electricity distribution utilities are very similar. Generally, the peak period of electricity consumption ranges from 6 AM to 10 PM. Moreover, some electric utilities consider part of the weekend as an off-peak period of electricity consumption.

PART I

ELECTRICITY DEMAND BY TIME-OF-USE

INTRODUCTION TO PART I

In this part we examine the residential electricity demand by time-of-use in Switzerland. The main purpose of this analysis is to estimate the price and expenditure elasticities of demand for peak and off-peak electricity. In addition to quantifying the impact of prices and expenditure, we identify the major household characteristics responsible for changing electricity market shares for peak and off-peak electricity.

In principle, a demand system for electricity consumption by time-of-use can be estimated by employing information on electricity expenditures, prices and household characteristics at aggregate levels, for example at the city or state level, or at micro levels, such as at the household or household group level or combining information pertaining to different levels of aggregation, individual households, cities or states.

To estimate price elasticities the aggregate data sets are generally more indicative than the micro data sets because they show a greater time series, cross section or panel variation in electricity price variables. The micro data sets usually refer to samples of households in the same city or state that are faced with the same prices. Therefore, employing micro data does not allow empirical identification of the parameters on electricity price variables which are of main interest in demand studies. However, aggregate data sets have the disadvantage of containing inaccurate information on household characteristics which are usually well documented by micro data sets. Thus, each data set incorporates useful information on some of the relevant variables for the estimation of a demand system for electricity, but none of them to permit correct estimates of all the coefficients of interest. Moreover, employing aggregate data sets gives rise to the problem of aggregation: how to specify a market demand equation that can be estimated with available data in a manner that is consistent with microeconomic theory so that the powerful tools of this theory may be employed.

In this study, we use three different data sets to estimate electricity demand by time-of-use.[12] The first data set contains information on electricity expenditures, prices and household characteristics at the aggregate level, for 21 Swiss cities covering the period 1987-1990. The second data set comprises data for the years 1989-1990 on electricity consumption and household characteristics for 280 households in the city of Zurich . While the third data set includes information covering the year 1991 for 217 households in 19 Swiss cities. The use of these data sets exibit the advantages and disadvantages presented above. The first data set does not contain accurate information on household characteristics but shows variations of the electricity price variable in space and time. The second data set contains detailed information about household characteristics but does not show variation in the price variables, while the the third data set contains precise information on household characteristics but shows a variation in the electricity price variables that is not extremely high.

To deal with the characteristics of these data sets we decided to analyze the residential demand for electricity by time-of-use in Switzerland using three approaches. The first approach consists of the application of the representative consumer model proposed by Muellbauer (1975, 1976). This model is estimated using the aggregate data set. The second approach is based on the aggregate consumer behavior model suggested by Jorgenson, Lau and Stoker (1982) and on the exact aggregation theory of Lau (1982). This model allows the simultaneous estimation of prices, expenditure and household characteristic effects by pooling the aggregate data set with the micro data set on households living in Zurich. Both models satisfy the conditions under which an aggregate data set should reflect individuals optimizing behavior. Whereas the third approach employs the micro data set

---

12 When we began this study in 1991 no data sets were available with all needed information for the estimation of Swiss residential electricity demand by time-of-use. All available aggregate or micro data sets had to be completed with information from our own survey. Therefore, at that time, we were not able to know exactly when and which type of data set would have had the most useful information for the purpose of our study. Thus, to have definitive data sets with useful information for estimation purposes, we decided to build three data sets.

for 217 households living in 19 other Swiss cities and estimates electricity demand by time-of-use as set for by the household consumer behavior model introduced by Deaton and Muellbauer (1980a).

The main objectives of estimating electricity demand by time-of-use using these three model specifications are as follows:

- To provide the policymaker with more information on the effects of price and household characteristic variables on the electricity consumption pattern;
- To address relevant model specifications and estimation issues for a system of demand equations given data sets pertaining to different levels of aggregation;
- To address the reliability and consistency of different model specifications in estimating system of demand equations using aggregate and micro data sets.

This part of the book is organized into five chapters. The next chapter discusses the problems and issues related to implementing consumer theory. The third chapter presents an empirical analysis of residential electricity demand by time-of-use using the representative consumer approach, while in chapter 4 the results of the estimation of the aggregate consumer behavior model are discussed. We then provide in chapter 5 the estimation results of the household consumer behavior model, whereas in chapter 6 a summary and conclusions are presented.

CHAPTER 2

IMPLEMENTING CONSUMER THEORY

Several elements of the theory of consumer behavior are introduced in this chapter which will proof helpful later on for the analysis of electricity demand. The first section reviews the basic theoretical approaches to generate a system of demand equations. In the second section we present the two-stage budgeting approach in the analysis of consumer behavior. The third section shows different models to estimate demand functions using aggregate data sets. Whereas in section four the welfare measurement problems within the two-stage budgeting modelling are presented. In section five we introduce the functional form utilized in modelling the Swiss residential electricity demand by time-of-use, while in the last section a survey of the most important previous studies on electricity demand by time-of-use are presented.

2.1 Review of consumer theory

The microeconomic theory of consumer demand leading to systems of demand functions is well documented in the literature, e.g., Deaton and Muellbauer (1980b), Thomas (1987), Jehle (1991), Varian (1992) and Pollak and Wales (1992), and it is not the purpose of this section to repeat this material in detail. Instead, this section focuses on some elements of the microeconomic theory of consumer behavior that are relevant for the generation of a demand system model to analyze the electricity demand by time-of-use in Switzerland.

There are three principal approaches to derive a demand system based on the microeconomic theory of consumer demand: the direct utility function approach, the indirect utility function approach, and the expenditure function approach.

Direct utility function approach

Neoclassical consumer theory attempts to explain how consumers make consumption choices at a given point in time. The basic elements of the theory include the utility function, the commodity set, and the axioms concerning the ordering of preferences. The utility function can be defined as an interpretative measure of the satisfaction derived from the consumption of alternative commodity bundles.

There are five axioms on consumer decision behavior that guarantee the existence of a utility function: non-satiation of wants, reflexivity, completeness, transitivity, and continuity. The basic implication of imposing these assumptions on preferences is that the rational consumer will always maximize utility from a given total expenditure (or income) level.

Formally, the consumer maximizes the utility function

$$u = u(\mathbf{q}) \tag{2.1}$$

subject to the linear budget constraint

$$\mathbf{p}'\,\mathbf{q} = m \tag{2.2}$$

where u refers to the utility function; $\mathbf{q}' = (q_1, q_2, ...,q_n)$ is the vector of quantities of commodities 1,2,...., n; $\mathbf{p}' = (p_1, p_2, ...,p_n)$ is the vector of prices of commodities 1,2,...., n; m is total expenditure or income.[13] The

13 If all income is spent in each time period, total expenditure and income are the same. However, a share of income is generally saved and this leads us to consider the consumer as an inter-temporal utility maximizer. Fortunately, assuming that preferences are weakly inter-temporally separable is sufficient to allow writing

utility function is assumed to be twice continuously differentiable, strictly increasing and strictly quasi-concave in the quantities.

The consumer's choice of individual commodities corresponds to the quantities consistent with the maximization of the following Lagrange function:

$$L = u(\mathbf{q}) + \lambda\,(m - \mathbf{p}'\,\mathbf{q}) \tag{2.3}$$

where λ is a Lagrange multiplier which is interpreted as the marginal utility of income. Necessary conditions for an interior solution are:

$$\frac{\partial u}{\partial q_i} - \lambda p_i = 0 \qquad (i=1,2,\ldots.,n)$$

and (2.4)

$$m - \mathbf{p}'\,\mathbf{q} = 0$$

The solution to the system (2.4) gives the n (Marshallian) demand functions and the equilibrium value of λ, as follows:

$$q_i = q_i(\mathbf{p}, m) \qquad (\mathrm{i} = 1,2,\ldots.,\mathrm{n}) \tag{2.5}$$

and

$$\lambda = \lambda\,(\mathbf{p}, m) \tag{2.6}$$

The n equations in (2.5) constitute the complete set of consumer demand functions, from which economic information on consumption behavior can be derived. The optimal quantity purchased of each commodity is expressed as a function of total expenditure and commodity prices.

---

demand equations as functions of current prices and current total expenditure. The weak separability assumption is discussed further in Section 2.2.

Indirect utility function approach

The indirect utility function expresses utility as a function of prices and total expenditure rather than quantities consumed directly. That is, the indirect utility function gives the maximum attainable utility for any given prices and total expenditure.

The indirect utility function can be obtained by substituting the demand functions (2.5) into the utility function (2.1). This substitution gives

$$\begin{aligned} u^* &= u(q_1^*(p_1, p_2, ...,p_n, m), ..., q_n^*(p_1, p_2, ...,p_n, m)) \\ &= v(\mathbf{p}, m) \end{aligned} \tag{2.7}$$

An indirect utility function corresponds to a well-behaved preference ordering if it is continuous, decreasing in prices, increasing in total expenditure, strictly quasi-convex in prices, and homogeneous of degree zero in prices and income.

The advantage to derive a demand system using the indirect utility function is that a system of (Marshallian) demand functions can be easily derived by differentiating (2.7) with respect to prices and total expenditure and applying Roy's identity:

$$\frac{-\left[\partial v(\mathbf{p},m)\right]}{\partial p_i} \Big/ \frac{\left[\partial v(\mathbf{p},m)\right]}{\partial m} = q_i(\mathbf{p},m) \tag{2.8}$$

Expenditure function approach

An alternative convenient formulation of the consumer's problem of choice is in terms of the expenditure function. The expenditure function indicates the minimum required expenditure at which a given utility u^* can be obtained at given prices $p_1, p_2, ...,p_n$. The expenditure function corresponding to the maximization problem (2.3) is stated here as:

$$e(\mathbf{p}, u^*) = \min \mathbf{p}'\mathbf{q}$$

$$\text{s.t. } u(\mathbf{q}) = u^* \tag{2.9}$$

where $e(\mathbf{p}, u^*)$ is the expenditure function. Therefore, the solution to the primal problem of maximizing utility for a given amount of total expenditure and prices produces a maximum attainable utility u^*. Whereas the dual problem is defined in (2.9) as the minimization of the expenditure necessary to reach the level of utility u^*.

The expenditure function has the following properties: homogeneous of degree one in prices, concave in prices, non-decreasing in prices and increasing in u. Additionally, differentiating $e(\mathbf{p}, u^*)$ with respect to prices leads to the Hicksian demand functions:

$$\frac{\partial e(\mathbf{p},u)}{\partial p_i} = q_i^h(\mathbf{p},u^*) \tag{2.10}$$

This property is commonly known as Shephard's lemma.

The primal and dual problems are alternative ways of representing the optimal consumer allocation choice. The theory of duality in consumer demand shows how the direct utility function, the indirect utility function, the expenditure function and the Hicksian and Marshallian demand functions are linked together. For example, using Shephard's lemma, Marshallian demand functions can be derived from the expenditure function by simple substitution. In the same way, Hicksian demand functions can be obtained by substituting the expenditure function in the Marshallian demand functions. Further, there is a close relationship between the indirect utility function and the expenditure function: they are simply the appropriately chosen inverses of each other.

Finally, the fundamental theorem of duality in consumer demand shows that any function $e(\mathbf{p}, u^*)$ that satisfies the properties presented above can be regarded as an expenditure function representing some underlying preference ordering.[14] This result allows applied researchers to recover from the estimation of a demand system derived from an expenditure or an indirect utility function economically relevant information about consumer preferences and choices without writing down the utility function. This fact simplifies applied consumption analysis, because

---

14 For a rigorous presentation of duality in consumer theory see Cornes (1992).

deriving a system of demand equations for empirical implementation from a utility function can be very inconvenient.[15]

Properties of demand functions

Demand functions, to be logically consistent with the underlying microeconomic theory of consumer behavior from which they are constructed, have to possess the following properties: adding-up, homogeneity, symmetry and negativity. Each of these properties defines an exact set of relationships which any complete set of demand functions must possess.

The first property of demand functions, adding-up, states that expenditures across commodities sum to total expenditures. This property requires that the following two aggregation conditions hold:

$$\sum_{i=1}^{n} w_i \eta_i = 1 \qquad i=1,....,n. \quad \text{(Engel aggregation)} \qquad (2.11)$$

$$\sum_{i=1}^{n} w_i \varepsilon_{ij} = - w_j \qquad j=1,....,n. \quad \text{(Cournot aggregation)} \quad (2.12)$$

The first condition is known as the Engel aggregation and states that the sum of the expenditure elasticities (η_i) weighted by their respective average budget shares w_i equals one. Whereas the second means that the weighted sum of the cross-price elasticities (ε_{ij}) of the jth commodity plus the direct price elasticity (ε_{jj}) of the jth commodity equals the negative of the average budget share for the jth commodity.

The second property, homogeneity, implies that demand functions are homogeneous of degree zero in prices and total expenditure. Therefore, if all prices and total expenditure change proportionally, the quantity

---

15 For a comparaison of the empirical performance of different consumer allocation models see Barten (1993).

demanded of commodities remain unchanged. In terms of price and income elasticities, the homogeneity property implies:

$$\sum_{j=1}^{n} \varepsilon_{ij} + \eta_i = 0 \qquad i=1,......,n. \tag{2.13}$$

The third property, symmetry, indicates the relationship among cross-substitution effects. Define the Slutzky matrix **S** by

$$\mathbf{S} = \begin{bmatrix} S_{11} \ .... \ S_{1n} \\ \vdots \\ S_{n1} \ .... S_{nn} \end{bmatrix} \tag{2.14}$$

where S_{ij} is the cross partial derivative of the compensated demands

$$S_{ij} = \frac{\partial q_i^h}{\partial p_j} \tag{2.15}$$

Each term of this matrix shows how quantity demanded for commodity i changes when the price of commodity j changes holding utility constant. Then, the symmetry property indicates that for all $i \neq j$, it is true that

$$S_{ij} = S_{ji} \tag{2.16}$$

therefore, the symmetry condition implies that compensated cross-price derivatives are equal.

The fourth property builds on the third. Negativity is satisfied if symmetry holds, and the Slutzky matrix **S** is negative semi-definite. In practical terms, it restricts the diagonal elements of the matrix **S**, which are the own compensated price terms, to be negative. Negativity implies that if the price of a commodity increases and the consumer is compensated so as to maintain utility at a constant level, then the quantity demanded for that good will decrease.

In general, from the view of applied consumption analysis there are two ways to formulate empirical models of complete demand systems.[16] The first method is to specify the functional form of the demand equations directly and then impose the general theoretical restrictions (adding-up, homogeneity, symmetry and negativity). The double-logarithmic demand system and the Rotterdam demand system are examples of applications of this first approach. The main shortcoming of this approach is that the functional forms of the demand equations are not adequate in reflecting the properties of demand functions.

The second method of deriving demand systems involves the specification of a direct utility, indirect utility or expenditure function. As presented above, the solution of the constrained utility maximization problem generates a set of demand equations that automatically satisfy the properties of demand functions, and whose functional form depends on the form of the utility function. The problem of this approach is that the number of well-behaved utility functions is very limited and the derivation of a demand system is not always possible. Moreover, in many cases, the derived demand functions are too complicated to permit estimation. Generally, these problems disappear when a system of demand equations is generated by differentiating a well-behaved indirect utility or expenditure function. Therefore, from an empirical point of view, the most attractive method of deriving a system of demand equations consists of specifying a functional form for the indirect utility function or the expenditure function and then applying Roy's identity or Shephard's lemma, respectively.

2.2 Two-stage budgeting approach

From a theoretical point of view there are no problems in formulating a complete demand system for a consumer faced with a multitude of goods among which to allocate his total current expenditure. However, from an empirical point of view the estimation of a complete demand system can involve serious problems. First, if the number of goods is large relative to sample size, the researcher would run out of the degrees

16 See Barten (1993)

of freedom when trying to estimate a complete demand system. Second, the estimation task, the cost of estimation and the problem to find a data set with all the necessary information for the estimation of complete demand systems can be considerable.

There are two basic ways by which it is possible to simplify these problems. One is to aggregate over certain groups of goods in such a way as to reduce the number of goods that need to be dealt with. The other is to represent the consumption decision of a consumer in stages. This involves the idea of two-stage budgeting process discussed by Strotz (1959) which is appealing and seems highly plausible. The first stage involves allocation of total expenditure to broad groups of goods, such as food and cars. The only information required at this stage is the knowledge of total expenditure to be allocated and appropriate price indices for each group. The second stages implie the optimal allocations of the group expenditures determined in the first stage to goods within each group. The information required at this stage is the group expenditures and the prices of goods in each group. It is clear that the first stage of this two-stage budgeting approach involves aggregation over certain groups of goods.

The main advantage of the two-stage budgeting process is that it allows the analyst interested in only a subset of goods to estimate just the second stage (within group allocation) of the two-stage budgeting process, reducing therefore the data and estimation problems presented above. However, for an exact two-stage budgeting process the results of the process must be identical to what would occur if the allocation were made in one step with complete information.[17]

In general notation the two-stage budgeting process can be written:

$$m_n = v_n\,(\,m, \mathbf{p}\,) \qquad n = 1,\ldots, \mathrm{N} \qquad (2.17)$$

$$x_{ng} = f_{ng}\,(\,m_n, \mathbf{p}_\mathrm{n}\,) \qquad g = 1,\ldots, \mathrm{G} \qquad (2.18)$$

---

17 See Deaton and Muellbauer (1980b).

where m_n is expenditure on group n, m is total expenditure, $\mathbf{p}$ is a vector of price indices of the groups, x_{ng} demand of good $g \in n$ and $\mathbf{p}_n$ is a vector of prices in group n. Thus, there are N groups of goods, each comprising g individual goods. The equations (2.18) are called conditional demand functions because the demand of good g depends on the preallocated total expenditure on group n. A conditional demand system is composed of g conditional demand functions.

The two-stage budgeting approach raises two questions. First, under which assumptions it is possible to express demand for any good g within a group n as a function of only total expenditure on group n and the prices within that group? Second, when is it possible to compress the influence of individual prices in group n into a single price index with which to determine, along with total expenditure and other prices, overall expenditure on group n; i.e., under which conditions, is the two-stage budgeting approach consistent with a one-stage procedure?

Focusing on the first question, if preferences are said to be weakly separable then it is possible to express the demand for any good g entirely in terms of expenditure on its group n and intra-group prices. A necessary and sufficient condition for weak separability is that the utility function is weakly separable over the groups $n = 1,....N$; namely

$$U = F\left(u_1(\mathbf{x}_1), u_2(\mathbf{x}_2), u_3(\mathbf{x}_3), ..., u_N(\mathbf{x}_N) \right) \tag{2.19}$$

where F is an increasing function of its arguments and $\mathbf{x}_n$ is a vector of goods, x_{ng}, in group n. The u_n are known as subutility functions. The utility functions of form (2.19) have the following three implications. First, the marginal rate of substitution between any two goods in the same group is independent of the quantity of any good outside the group. Mathematically,

$$\frac{\partial(u_i / u_j)}{\partial x_k} = 0 \tag{2.20}$$

for all i, j n, and k s, $n \neq s$. Second, total expenditure and the prices of goods outside the commodity group affect the demand of good i n, x_i

only through their effect on the expenditure on that group. Formally, weak separability implies:

$$\left.\frac{\partial x_i}{\partial p_k}\right|_{u\ const.} = \frac{\partial x_i}{\partial m_n}\frac{\partial m_n}{\partial p_k}, \qquad i \in n,\ k \in s,\ n \neq s$$

$$\left.\frac{\partial x_i}{\partial m}\right|_{u\ const.} = \frac{\partial x_i}{\partial m_n}\frac{\partial m_n}{\partial m}, \qquad i \in n. \tag{2.21}$$

Moreover, a change in the price of good k or a change in total expenditure has the same effect on the demand for good x_i.

The two-stage budgeting process has to be consistent with a one-stage procedure, i.e., the quantity of each good determined by the two-stage process has to be identical with the amount which would have been chosen if utility had been maximized with reference to all the individual prices, without a grouping process. In general, following Deaton and Muellbauer (1980b) when there exist more than two groups the necessary and sufficient conditions for the consistency of the two-stage maximization procedure are:

(a) the utility function be weakly separable;

(b1) the expenditure functions have the form:

$$e_n(u_n, \mathbf{p}_n) = \theta_n(u_n) b_n(\mathbf{p}_n) \tag{2.22}$$

where $\theta_n(.)$ is a monotonic increasing function, $b_n(.)$ is a linear homogeneous function, and u_n is the sub-utility level for which the cost function is defined. This particular form of the expenditure function implies that the sub-utility functions are homothetic, i.e., the within group budget shares are independent of the level of total expenditure on that group, **or**

(b2) the overall utility function is explicitly additive, while the indirect utility function takes the Gorman generalized polar form

$$U = F\,(\,u_1(\mathbf{x_1}) + u_2(\mathbf{x_2}) + u_3(\mathbf{x_3}) + ... + u_N(\mathbf{x_N})\,) \tag{2.23}$$

and

$$y_n(\mathbf{x}_n, \mathbf{p}_n) = F_n(\mathbf{x}_n / b_n(\mathbf{p}_n)) + a_n(\mathbf{p_n}) = u_n \tag{2.24}$$

Both conditions (b1) and (b2) are generally considered quite restrictive. The first implies unitary elasticities with respect to the total within group expenditure. Whereas the second assumes an additive utility function which implies that inferior goods are ruled out and that the goods are allowed to be substitutes only. However, for the analyst interested in only the estimation of a conditional demand system, the technical requirement imposed by condition (b2) is more attractive than that imposed by (b1), because, in this case, the analyst estimates just the second stage (within group allocation) of the two-stage budgeting process, avoiding the empirical implementation of the restrictive first stage of the budgeting process.

Finally, although the two-stage budgeting procedure implies some restrictive assumptions it has the major advantage of reducing the data required to analyze demand within-groups and therefore, to allow the empirical estimation of economic and demographic effects for demand of numerous goods.

2.3 Aggregation and consumer behavior

The micro-theory of consumer behavior is derived from postulates about optimizing behavior on the part of individuals. Unfortunately, data on prices and quantities purchased frequently referr to aggregate groups of consumers. This creates problems, first because the demand system estimated with aggregate data will not necessarily reflect individual optimizing behavior, and second because the model for individual consumer behavior does not in general aggregate to a convenient representation of aggregate consumer behavior. In this view, the role of aggregation theory is to provide the necessary conditions under which aggregate demands can be treated as if they were the outcome of the decisions of a single maximizing consumer, i.e., the theoretical conditions under which it is possible to estimate the parameters of micro demand behavior from aggregate data. In what follows, for information purposes only, three different approaches to aggregation over consumers are briefly presented: exact linear aggregation, exact non-linear aggregation and Lau's theory of exact aggregation.[18]

Exact linear aggregation

In a given population there are $h = 1, \ldots, H$ consumers each with demand functions,

$$q_{hi} = f_{ih}(m_h, \mathbf{p}) \tag{2.25}$$

where q_{hi} denotes the demand for good i of consumer h, m_h is total expenditure of consumer h, and $\mathbf{p}$ is a price vector. Using aggregate data, we observe only the sum of the individual consumer demands:

$$q_i = \sum_{h=1}^{H} f_{ih}(m_h, \mathbf{p}) \tag{2.26}$$

therefore, the average demand $\bar{q}_i$ will be

$$\bar{q}_i = \frac{1}{H} \sum_{h=1}^{H} f_{ih}(m_h, \mathbf{p}) \tag{2.27}$$

total expenditure varies across consumers, but prices do not. Exact aggregate demand functions exist if it is possible to write the average demand $\bar{q}_i$ in the following form, for some

f_i, and for all i :

$$\bar{q}_i = f_i(\bar{m}, \mathbf{p}) \tag{2.28}$$

---

18 For a detailed presentation of the aggregation issues in consumer theory see Deaton and Muellbauer (1980b), Jorgenson et al. (1982), Thomas (1987) and Stoker (1993).

where $\overline{m}$ is average total expenditure and the average demand $\overline{q}_i$ is a function of all consumers total expenditure and prices. Note that (2.27) is not necessarily equal to (2.28). For equation (2.28) to hold, a change in the expenditure distribution should not affect aggregate demands and therefore average demands. This occurs if all consumers have the same marginal propensities to consume for any good. This implies that all individual demands must be linear in consumer total expenditure m_h and have the same slope with respect to m_h. Generally the individual demand equations must take the following form

$$q_{hi} = \alpha_{hi}(\mathbf{p}) + \beta_i(\mathbf{p}) m_h \tag{2.29}$$

so that we can aggregate and obtain

$$\overline{q}_i = \alpha_i(\mathbf{p}) + \beta_i(\mathbf{p})\overline{m} \tag{2.30}$$

where $\alpha_i(\mathbf{p})$ is the average of the $\alpha_{hi}(\mathbf{p})$, and average demand does not depend on the distribution of individual expenditures m_h. In this case, Engel curves are linear.

If individuals are assumed to maximize utility, Gorman (1953, 1961) showed that (2.29) holds if and only if the individual expenditure functions have the form:

$$e_h(u_h, \mathbf{p}) = a_h(\mathbf{p}) + b(\mathbf{p}) u_h \tag{2.31}$$

This form of the expenditure function, which represent quasi-homothetic preferences, is called the Gorman Polar form, because it was Gorman (1953) who first suggested it. The functions $a_h(.)$ and $b(.)$ have an intuitive interpretation; $a_h(.)$ is the expenditure of attaining the standard of living corresponding to a utility level of zero, and $b(.)$ is the additional expenditure necessary to attain utility level u_h.. Moreover, to ensure that (2.31) possesses all the properties of a well behaved consumer cost function, it is necessary and sufficient that the functions $a_h(.)$ and $b(.)$ be concave in prices.

Summing over all h and dividing by the total number of consumers yields the aggregate expenditure function:

$$\overline{m} = e(u,\mathbf{p}) = a(\mathbf{p}) + b(\mathbf{p})\overline{u} \tag{2.32}$$

where $\overline{m}$ is mean expenditure and $\overline{u}$ is mean utility. The preferences referred to in (2.32) can be thought of as those of an arithmetic mean consumer. Therefore, the aggregate equations (2.28) can be regarded as if they result from the utility maximization process of an arithmetic mean consumer.

In conclusion, quasi-homothetic preferences or, equivalently, linear Engel curves are necessary conditions for exact linear aggregation.

Exact non-linear aggregation

The conditions for exact linear aggregation are very restrictive. However, Muellbauer (1975, 1976) developed a necessary and sufficient condition which allows exact aggregation with non-linear Engel curves.[19] Muellbauer showed that, provided the individual expenditure functions assume a certain form, then it is possible to aggregate individual budget share equations into a macro equation which can be thought of as deriving from the expenditure function of some representative utility maximizing consumer. The aggregate budget shares depend on prices and a representative level of total expenditure, which itself can be a function of the distribution of expenditures and prices. Thus, Muellbauer defines exact aggregation over budget shares rather than quantities as in the case of linear exact aggregation.

A useful special case, known as price independent generalized linearity (PIGL), arises when the representative expenditure is independent of prices and depends only on the distribution of individual expenditures.

To achieve exact non-linear aggregation in this case, Muellbauer's analysis shows that the expenditure function of each household have to take the form:

---

19 For a detailed presentation of the nonlinear exact aggregation see Muellbauer (1975, 1976) and Cornes (1992).

$$e_h(u_h,\mathbf{p}) = k_h\left[a(\mathbf{p})^{\alpha}(1-u_h)+b(\mathbf{p})^{\alpha}u_h\right]^{\frac{1}{\alpha}} \tag{2.33}$$

where α, and k^h are parameters, and $b(.)$ and $a(.)$ are linearly homogeneous in prices. Note that the cost function (2.33) is a generalization of the Gorman Polar form cost function (2.31) in which the expenditure level is a constant eleasticity of substitution function (CES) of $b(.)$ and $a(.)$.

The expenditure function of the representative household must be

$$e_r(u_r,\mathbf{p}) = \left[a(\mathbf{p})^{\alpha}(1-u_r)+b(\mathbf{p})^{\alpha}u_r\right]^{\frac{1}{\alpha}} \tag{2.34}$$

The parameters α determines the degree of non-linearity of the Engel curves and therefore the relationship between representative and average expenditures as well. For example, when α equals one, (2.34) takes the linear form and the Engel curves are linear. Further, it may be shown that when α tends to zero (2.34) becomes:

$$e_r(u_r,\mathbf{p}) = [a(\mathbf{p})]^{1-u_r}+[b(\mathbf{p})]^{u_r} \tag{2.35}$$

A special case of PIGL is the PIGLOG form whose individual expenditure function is

$$\ln e_h(u_h,\mathbf{p}) = (1-u_h)\ln a(\mathbf{p})+u_h\ln b(\mathbf{p}) \tag{2.36}$$

The expenditure function of the representative consumer must be

$$\ln e_r(u_r,\mathbf{p}) = (1-u_r)\ln a(\mathbf{p})+u_r\ln b(\mathbf{p}) \tag{2.37}$$

Muellbauer's work shows that exact aggregation is possible without the requirement of linear Engel curves within the demand system generated by the PIGL family. Therefore, in Muellbauer's model of the representative consumer, individual preferences are the same but not necessarily homothetic.

The Almost Ideal Demand System (AIDS), employed in the following chapters for the empirical analysis of the electricity demand by time-of-use, belongs to the PIGLOG family, and thus, it adds up exactly over households.[20]

Lau's theory of exact aggregation

Lau (1982) proposed a theory of exact aggregation which is more general than the traditional linear and non-linear aggregation approaches proposed by Gorman (1953) and Muellbauer (1975, 1976).

The generalization introduced by Lau's theory of exact aggregation consists of allowing aggregate demand, obtained by exact aggregation over individual demand functions, to depend on prices and on general symmetric functions of individual expenditure and attributes (e.g., demographic characteristics). Therefore, by allowing the aggregate quantities demanded to depend on these symmetric functions, Lau developed a theory of exact aggregation that makes it possible to incorporate differences in individual preferences and thus overcomes the limitation of the model of the representative consumer.

He considered individual demand function of the form:

$$q_{hi} = f_{hi}(m_h, \mathbf{p}, \mathbf{c}_h) \qquad i = 1,2,\ldots,I \quad and \quad h = 1,2,\ldots,H. \tag{2.38}$$

where q_{hi} denote the demand for good i of consumer h, m_h is total expenditure of the hth consumer, $\mathbf{p}$ is a price vector of all goods and $\mathbf{c}_h$ is a vector of characteristics for the hth consumer. The aggregate demand function becomes:

$$\begin{aligned} \sum_h f_{hi}(m_h, \mathbf{p}, \mathbf{c}_h) = F_i(\mathbf{p}, G_1(m_1, m_2, \ldots, m_H, \mathbf{c}_1, \mathbf{c}_2, \ldots, \mathbf{c}_H), \ldots, \\ G_K(m_1, m_2, \ldots, m_H, \mathbf{c}_1, \mathbf{c}_2, \ldots, \mathbf{c}_H)) \qquad G_k = 1, \ldots\ldots, K \end{aligned} \tag{2.39}$$

20 See Deaton and Muellbauer (1980a).

where F_i is a function giving the aggregate demand for good i and G_k is a non-constant symmetric function of the m_h and $\mathbf{c}_h$. Symmetry of the functions G_k means that the labelling of individuals is not important. Lau refers to the functions G_k as index functions, which can be interpreted as statistics describing the population.

The Fundamental Theorem of Exact Aggregation (see Lau,1982) gives the conditions under which one can write an aggregate demand function in the form of (2.39).[21] From an economic point of view this theorem has the following striking implications:

1. individual demand functions are the same for all consumers with the same total expenditure and individual characteristics except for an additive function of price alone;

2. individual demand functions must be sums of products of separate functions of the prices and of the individual expenditures and characteristics;

3. the only admissible index functions are additive in functions of individual expenditures and characteristics;

4. aggregate demand functions can be written as linear functions of the index functions.

The first two implications indicate that the only admissible forms for the individual demands are

$$f_h(m_h, \mathbf{p}, \mathbf{c}_h) = b_h^0(\mathbf{p}) + \sum_{k=1}^{K} b_k(\mathbf{p}) g_k(m_h, \mathbf{c}_h) \qquad (2.40)$$

where $b_h^0(\mathbf{p})$ is an additive function of price and $b_k(\mathbf{p}) g_k(m_h, \mathbf{c}_h)$ is the product of price-dependent and household-specific functions. Whereas

---

21 For a mathematical presentation of the exact hypotheses and conditions see Lau (1982) and Jorgenson et al. (1982) and Heineke (1993).

implication 3 states that the admissible index functions must take the following form

$$G_k(m_1, m_2, \ldots, m_H, \mathbf{c}_1, \mathbf{c}_2, \ldots, \mathbf{c}_H) = \sum_{h=1}^{H} g_k(m_h, \mathbf{c}_h) \tag{2.41}$$

This excludes index functions that are not additive in functions of individual expenditures and characteristics.

One of the most interesting implications of the model of aggregate consumer behavior based on Lau's theory of exact aggregation is that a system of demand functions for an individual with given demographic characteristics can be recovered uniquely from the system of aggregate demand functions. Therefore, this model of aggregate consumer behavior can be implemented from time-series, cross-section or panel aggregate data or by pooling this aggregate data set with a micro data set. As we will present in chapter 4, this pooling approach is appealing when the problem is to identify price effects on demand when the disaggregated data set does not show price variation, while the aggregate one does.

As shown by Rossi (1988) and Browning (1993) the Almost Ideal Demand System satisfies the conditions for exact aggregation and will be estimated in chapter 4 by pooling aggregate panel data with micro cross section data.

2.4 Compensating and equivalent variation in a conditional demand system

Assume that a consumer maximizes the utility function

$u_h = u_h(\mathbf{q}, \mathbf{s})$ subject to the linear budget constraint

$$\mathbf{p}'\mathbf{q} + \mathbf{r}'\mathbf{s} = m \tag{2.42}$$

where u_h refers to the utility function of household h, $h = 1,2,\ldots, H$; $\mathbf{q}' = (q_1, q_2, \ldots, q_n)$ is the vector of quantities for a subset of commodities; $\mathbf{s}' = (s_1, s_2, \ldots, s_k)$ is the vector of quantities for all other commodities; $\mathbf{p}' = (p_1, p_2, \ldots, p_n)$ is the vector of prices of the subset of commodities, e.g. electricity commodities; $\mathbf{r} = (r_1, r_2, \ldots, r_k)$ is the vector of prices of all other commodities; m is total expenditure. The utility function is assumed to be twice continuously differentiable, strictly increasing and strictly quasi-concave in its quantities.

The expenditure function corresponding to the maximization problem (2.42) is stated here as:

$$e_h(\mathbf{p}, \mathbf{r}, u_h^*) = \min \mathbf{p}'\mathbf{q} + \mathbf{r}'\mathbf{s}$$

$$\text{s.t. } u(\mathbf{q}, \mathbf{s}) = u_h^* \qquad (2.43)$$

where $e_h(\mathbf{p}, \mathbf{r}, u_h^*)$ is the expenditure function of household h. This function can be used to define monetary measures of the welfare effects of changes in commodity prices. Although alternative measures have been proposed in the literature, for the purposes of this analysis we are going to use the concepts of compensating and equivalent variation.[22]

Compensating variation (CV) is the minimum amount of money by which a consumer would have to be compensated after a price change in order to be as well off as before. Formally, the general compensating variation measure of welfare for a household h following a change in the subset of commodity price vector $\mathbf{p}$, is

$$CV = e_h(\mathbf{p}^1, \mathbf{r}^0, u_h^0) - e_h(\mathbf{p}^0, \mathbf{r}^0, u_h^0) \qquad (2.44)$$

where $e_h(\mathbf{p}^1, \mathbf{r}^0, u_h^0)$ is the minimum expenditure of household h to achieve utility u_h^0 when facing the "policy case" price vector for the commodities under study $\mathbf{p}^1$ and $e_h(\mathbf{p}^0, \mathbf{r}^0, u_h^0)$ is the minimum expenditure of

22 The reader interested in other, less frequently employed welfare measures is referred to McKenzie (1983).

household h to achieve utility u_h^0 when facing the "base case" price vector $\mathbf{p}^0$. For instance, if the price of a subset commodity increases, CV is the minimum amount of money that must be given to the household to compensate it for that price increase. Therefore, the CV assumes there has been a move from "base case" price vector to "policy case" price vector.

Equivalent variation (EV) is the amount of money we would need to give the consumer, if an economic change did not happen, to make him as well off as if it did. Formally, the equivalent variation measure of welfare for a household h following a change in the commodity price vector $\mathbf{p}$, is

$$EV = e_h(\mathbf{p}^1, \mathbf{r}^0, u_h^1) - e_h(\mathbf{p}^0, \mathbf{r}^0, u_h^1) \tag{2.45}$$

where $e_h(\mathbf{p}^1, \mathbf{r}^0, u_h^1)$ is the minimum expenditure of household h to achieve utility u_h^1 when facing the "policy case" price vector $\mathbf{p}^1$ and $e_h(\mathbf{p}^0, \mathbf{r}^0, u_h^1)$ is the minimum expenditure of household h to achieve the new utility level u_h^1 when facing the "base case" price vector $\mathbf{p}^0$. For example, if the price of a subset commodity increases, EV is the maximum the household is willing to pay to prevent that price increase. Therefore, EV assumes there has been no move from "base case" price vector to "policy case" price vector.

Since expenditure is an increasing function of utility, the two measures of welfare presented are "money-metrics" for utility changes between the two cases.[23]

To compute the CV and EV associated with a change in the price of a subset commodity, the standard approach is to estimate a complete demand system derived from an expenditure or an indirect utility function; then, using the estimation results of this complete demand system, recover the coefficients of the expenditure function or of the indirect utility function and thus, calculate CV and EV.

However, as already presented in section 2.2, because of the difficulty in obtaining data sets with all necessary information regarding $\mathbf{p}$,

23 See Samuelson (1974) and Varian (1992) for a detailed presentation of the money metric utility notion.

q, **r**, and **s**, the estimation of a complete demand system cannot usually be performed. Therefore, in these cases the computation of CV and EV defined in (2.44) and (2.45) is not possible. A common solution to this problem is to assume a two-stage budgeting process, and to estimate a conditional demand system for a subset of commodities for which information is available. This conditional demand system is used to derive a conditional expenditure or a conditional indirect utility function, which can be used to define monetary measures of the welfare effects of changes in commodity prices.

The conditional compensating variation (CV_c) measure of welfare for a household h following a change in the commodity price vector **p**, is

$$CV_c = e_{qh}(\mathbf{p}^1, u_{qh}^0) - e_{qh}(\mathbf{p}^0, u_{qh}^0) \tag{2.46}$$

where $e_{qh}(\mathbf{p}^1, u_{qh}^0)$ is the new group expenditure of household h needed to achieve utility u_{qh}^0 from the consumption of the subset of commodities g when facing the "policy case" price vector $\mathbf{p}^1$ and $e_{qh}(\mathbf{p}^0, u_{qh}^0)$ is the optimal allocation of expenditure to the subset of commodities of household h to achieve utility u_{qh}^0 when facing the "base case" price vector $\mathbf{p}^0$, which is determined in the first stage of the two stage budgeting process. This conditional welfare measure considers a situation where the household after the price change from $\mathbf{p}^0$ to $\mathbf{p}^1$ is restricted to continue to consume $\mathbf{s}^0$.

The conditional equivalent variation (EV_c) measure of welfare for a household h following a change in the commodity price vector **p**, is

$$EV_c = e_{qh}(\mathbf{p}^1, u_{qh}^1) - e_{gh}(\mathbf{p}^0, u_{qh}^1) \tag{2.47}$$

where $e_{qh}(\mathbf{p}^0, u_{qh}^1)$ is the group expenditure of household h needed to achieve the new utility level u_{qh}^1 when facing the price vector $\mathbf{p}^0$ and $e_{qh}(\mathbf{p}^1, u_{qh}^1)$ is the group expenditure of household h needed to achieve utility u_{qh}^1 when facing the new price vector $\mathbf{p}^1$.

The definition of conditional compensating and equivalent variations give rise to the following important question: when do EV_c and CV_c coincide with EV and CV?

The conditional compensating and equivalent variations treat expenditure for the subset of commodities as fixed either at the optimal level associated with the "base case" or the "policy case" prices. As show by Hanemann and Morey (1992) and LaFrance (1993) the two measures coincide only if the optimal allocation of expenditure to the subset of commodities in the current situation coincide with what would be the optimal allocation of expenditure to the subset of commodities in the "policy case" state. However, in the two-stage budgeting approach an increase in any of the prices of the subset commodities will cause an increase in the price index for the subset of commodities and therefore, a change in the optimal amount of expenditure to the subset of commodities. This implies that EV_c and CV_c will coincide with EV and CV only when the increase in the price index of the subset of commodities does not change the allocation of total expenditure among the different subsets of commodities. This situation occurs only when substitution between the subset and the remaining commodities is impossible. Therefore, EV_c and CV_c do not account for how easily the household substitutes between the **q** and **s** commodities.

Hanemann and Morey (1992) show that CV_c is only a lower bound on the desired CV. Furthermore, they show that EV_c is either a lower or a higher bound of EV.

The implication for the applied economist who estimates a conditional demand system for a subset of commodities and wants to estimate the effects on the consumer welfare from changes in prices is that CV_c understates CV. Therefore, the results obtained by calculating CV_c are biased and have to be interpreted carefully. LaFrance (1993) estimating both a conditional and a complete demand system found a difference of 17 % between CV_c and CV. This suggests that standard practice to measure welfare changes in an applied consumer analysis produces biased results.[24]

In the electricity demand by time-of-use literature we find several studies analyzing the consumer welfare effects of price changes.[25] However, none of them estimate welfare changes using a complete demand

---

24 See for example the studies of Blisard and Blaylock (1991) and Heien and Pompelli (1989).

25 For a presentation of the results of these studies see Aigner (1984).

system. For instance, Caves, Christensen, Schoech and Hendricks (1984) and Parks and Weitzel (1984) estimate consumer welfare effects of electricity price increases from an incomplete demand system, i.e., a subset of demand functions in a complete demand system. They assume a two-stage budgeting process where for the first stage a demand function for the total electricity consumption is estimated; and in stage II a homothetic conditional demand system for electricity demand by time-of-use is specified. The estimation results of the total electricity demand function were utilized to estimate welfare effects of electricity price changes. Nevertheless, as shown by Hausmann (1981) and by LaFrance and Hanemann (1989), the estimation results of an incomplete demand system do not contain sufficient information to correctly estimate *CV* and *EV*. Therefore, the effects upon consumer welfare reported in those studies should be interpreted carefully. Moreover, in a cost-benefit analysis conditional welfare measures can lead to wrong project evaluation. For these reasons, we decided to not perform an analysis of consumer welfare effects of time-of-use pricing for residential customers.

2.5 Functional form and the AIDS model

When applying consumer demand theory, selection of functional form is important. The two more common forms employed in applied consumer theory are linear and log-linear specifications. However, in the last two decades, with the advances brought by duality theory, many researchers have proposed particular indirect utility or expenditure functions using several flexible functional forms. These particular forms approximate the indirect utility function or the expenditure function by certain specific functional forms that have enough coefficients to be considered a reasonable approximation to the true unknown function.

Stern (1986) and Barten (1993) indicate the following criteria for evaluating functional forms for the estimation of a demand system: (1) easy of estimation, (2) adequacy in reflecting the theoretical restrictions of consumer theory, and (3) flexibility in the range of responses permitted.

As presented in section 2.2, given data limitations, the estimation of a complete demand system is not possible. Moreover, the micro consumption data are not always available. We have seen in section 2.2. and 2.3 that a solution to these data problems, that are common in electricity demand analysis, consists of the estimation of a conditional demand system using aggregate consumption data. Therefore, in addition to criteria (1), (2) and (3) we suggest considering the following two criteria: (4) consistency with a theoretical two-stage budgeting process, and (5) consistency with the exact aggregation over consumers.[26]

Table 2.1 presents some common functional forms employed in studies on electricity demand by time-of-use. Of the four functional forms presented in Table 2.1, the first one seems to be the least attractive because it needs severe restrictions on elasticities to be consistent with consumer demand theory. The second and the third approaches are not easy to estimate and are not consistent with the first stage of a two-stage budgeting process, although homotheticity is not imposed. Moreover, they do not aggregate exactly over consumers. Finally, the Almost Ideal Demand System (AIDS) expenditure function seems to be the most useful functional form when estimating a conditional demand system with aggregate data.[27]

26 As presented above, the necessary and sufficient conditions for the budgeting procedure to be consistent with a one stage procedure are that the group utility functions be homothetic or that the indirect utility functions have the generalized Gorman polar form and that the overall utility function is additive.

27 For a more detailed presentation of flexible functional forms see Chalfant (1984) and Barten (1993).

Table 2.1 Some common functional forms employed in studies on the electricity demand by time-of-use

Function	*Functional form*	*restrictions of economic theory are easily imposed*	*simple to estimate*	*aggregates perfectly over consumers*	*permits flexibility with regard to response*	*the indirect utility function takes the Gorman generalized polar form*
log-linear demand function	$ln\, E_i = \alpha_i + \sum_j \gamma_{ij}\, ln\, p_j + \beta_i\, ln\, m$	no	yes	no	no	no
Budget share equations derived from a homothetic translog indirect utility function	$S_i = \dfrac{\alpha_i + \sum_j \beta_{ij}\, ln(p_i / m)}{1 + \sum_k \sum_j \beta_{kj}(p_j / m)}$	yes	no, because it is a non-linear system	no	yes	no
Budget share equations derived from a generalized Leontieff indirect utility function	$S_i = \dfrac{\alpha_i (p_i / m)^{\frac{1}{2}} + \sum_j \beta_{ij}(p_i / m)^{\frac{1}{2}}(p_j / m)^{\frac{1}{2}}}{\sum_i \alpha_i (p_i / m)^{\frac{1}{2}} + \sum_j \sum_k \beta_{jk}(p_k / m)^{\frac{1}{2}}(p_j / m)^{\frac{1}{2}}}$	yes	no, because it is a non-linear system	no	yes	no
Budget share equations derived from the Almost Ideal Demand System expenditure function	$S_i = \alpha_i + \sum_j \gamma_{ij}\, ln\, p_j + \beta_i\, ln(m / P)$	yes	yes in its approxi-mate linear form	yes	yes	yes

E_i = demand of electricity commodities i
S_i = budget share of electricity commodities i
p_i = price of electricity commodities i
m = expenditures
P = price index

For this study, which estimates a conditional demand system for electricity and employs data at aggregate and micro levels, the Almost Ideal Demand System (AIDS), developed by Deaton and Muellbauer, has been selected.[28] This particular demand system, which is widely used in applied demand analysis, has a number of advantages. First, it is derived explicitly from utility maximization; second, as presented in section 2.3 it is suitable for the analysis of aggregate demand;[29] third, it is simple to estimate because it avoids non-linear estimation; fourth, the restrictions of economic theory (homogeneity and symmetry) are easily imposed. Fifth, elasticities need not be fixed parameters but can be made to depend on prices. Finally, employing the AIDS model, which is a special case of the generalized Gorman polar form, and implicitly assuming an additive overall utility function, the generated conditional demand system for electricity consumption by time-of-use is consistent with a one-stage procedure.

To derive the almost ideal demand system model, let the expenditure function take the following form:

$$ln\, e_h(u_h, \mathbf{p}) = (1-u_h)\, ln\, a(\mathbf{p}) + u_h\, ln\, b(\mathbf{p}) \tag{2.48}$$

where u_h denotes utility, $\mathbf{p}$ is a vector of commodity prices, and $a(\mathbf{p})$ and $b(\mathbf{p})$ are functions of prices alone that have to be chosen. To generate the almost ideal demand system, Deaton and Muellbauer (1980a) chose the following flexible functional forms for $a(\mathbf{p})$ and $b(\mathbf{p})$:

$$ln\, a(\mathbf{p}) = \alpha_0 + \sum_i \alpha_i\, ln\, p_i + \frac{1}{2}\sum_i \sum_j \gamma^*_{ij}\, ln\, p_i\, ln\, p_j \tag{2.49}$$

and

$$ln\, b(\mathbf{p}) = ln\, a(\mathbf{p}) + \beta_0 \prod_j p_j^{\beta_j} \tag{2.50}$$

---

28 See Deaton and Muellbauer (1980a).

29 The AIDS model for the analysis of aggregate demand is based on the aggregation theory developed by Muellbauer (1975, 1976).

Substituting these expression into (2.48) gives the expenditure function:

$$\ln e_h(u_h, \mathbf{p}) = \alpha_0 + \sum_i \alpha_i \ln p_i + \frac{1}{2}\sum_i\sum_j \gamma^*_{ij} \ln p_i \ln p_j + u_h \beta_0 \prod_j p_j^{\beta_j} \tag{2.51}$$

where α_i, γ^*_{ij} and β_i are parameters. e_h (.) is linearly homogeneous in prices as long as:

$$\sum_i \alpha_i = 1 \text{ and } \sum_i \gamma^*_{ij} = \sum_j \gamma^*_{ij} = \sum_j \beta_j = 0$$

By Shephard's lemma, expressions for budget shares are obtained as

$$w_{ih} = \alpha_i + \sum_j \gamma_{ji} \ln p_j + \beta_i u_h \beta_0 \prod_j p_j^{\beta_j} \qquad i, j = 1, \ldots\ldots, n$$

with $\gamma_{ij} = (\gamma^*_{ij} + \gamma^*_{ji})/2$. (2.52)

Since utility u_h is unobservable the budget share equations (2.52) cannot be estimated. By inverting the expenditure function (2.51) to obtain an expression for u_h and by substituting this expression in the budget share equations (2.52) we obtain

$$w_{ih} = \alpha_i + \sum_j \gamma_{ji} \ln p_j + \beta_i \ln(\frac{m_h}{P}) \qquad i, j = 1, \ldots\ldots, n. \tag{2.53}$$

where w_{ih} is the budget share of the ith good for the hth consumer, p_j are prices, and (m_h / P) is the total real expenditure on all n goods in the consumer's budget, with P a price index defined by

$$ln\, P = \alpha_0 + \sum_j \alpha_j \, ln\, p_j + \frac{1}{2}\sum_i \sum_j \gamma_{ij} \, ln\, p_i \, ln\, p_j \qquad (2.54)$$

To satisfy the properties of demand functions known as adding-up, homogeneity and symmetry, the following restrictions are imposed:

Adding up $$\sum_i \alpha_i = 1,\ \sum_i \gamma_{ij} = 0,\ \sum_i \beta_i = 0 \qquad (2.55)$$

Homogeneity $$\sum_j \gamma_{ij} = 0 \qquad (2.56)$$

Symmetry $$\gamma_{ij} = \gamma_{ji} \qquad (2.57)$$

The negativity condition, which is satisfied if the Allen-Uzawa Elasticities of Substitution (AUES) matrix is negative semi-definite, cannot be parametrically imposed in the AIDS model, but can be tested by evaluating AUES at some data point.[30]

Using the price index as defined in equation (2.54), the system of equations (2.53) becomes non-linear and requires the estimation of a large number of parameters. To linearize the AIDS model, we follow Deaton and Muellbauer (1980a) in their use of Stone's (geometric) price index as a linear approximation:

$$ln\, P \cong \sum_i w_i \, ln\, p_i \qquad (2.58)$$

The AIDS model using Stone's index is termed the "linear approximate almost ideal demand system".

30 The attribute "almost" in the name of the demand system (Almost Ideal Demand System) is attributable to the fact that the negativity condition cannot be guaranteed.

The calculations of the expenditure, substitution and uncompensated own- and cross-price elasticities used in this research follow Chalfant (1987).[31] The price elasticities for the linear approximate AIDS can be calculated, conditional on the price index chosen, as

$$\varepsilon_{ii} = -1 + \gamma_{ii} / w_i - \beta_i \tag{2.59}$$

$$\varepsilon_{ij} = \gamma_{ij} / w_i - \beta_i (w_j / w_i) \tag{2.60}$$

The expenditure elasticities are obtained from:

$$\eta_i = 1 + \beta_i / w_i \tag{2.61}$$

Using the Slutzky equation it is possible to show the relationship between the price, substitution, and expenditure elasticities.[32] The equation is given by

$$\varepsilon_{ij} = w_j \sigma_{ij} - w_j \eta_i \tag{2.62}$$

where σ_{ij} are the Allen elasticities of substitution. In the case of the the almost ideal demand system, the elasticities are obtained from

$$\sigma_{ij} = 1 + \gamma_{ij} / w_i w_j \qquad i \neq j \tag{2.63}$$

Expression (2.63) is very important because the usual definitions of complementary and substitute goods are based on the sign of σ_{ij}.

---

31 For a discussion on different approaches to calculate demand elasticities in AIDS see Alston and Greene (1990) and Alston, Foster and Green (1994).

32 In this case the income elasticity is replaced by the expenditure elasticity.

2.6 Review of the literature

For several decades, most of the European electric utilities have been applying time-of-use electricity rates in order to reflect peak-load cost variations and to promote efficient utilization of production and distribution capacities. By contrast, U.S. electric utilities began to consider the introduction of time-of-use pricing only during the 1970s, spurred in large part by the "energy crisis". Thus, the lack of experience with time-of-use rates on the part of U.S. utilities and the lack of statistical information about electricity consumption by time-of-use has resulted in a lack of empirical analysis of time-of-use electricity demand. As a result, the econometric literature on time-of-use demand consists of only a few studies.[33] This statement holds true in particular for residential demand for electricity.

In this section we present and assess some recent studies on the electricity demand by time-of-use, which we considered helpful in understanding the empirical analysis of the residential electricity demand by time-of-use in Switzerland presented in the next three chapters of this book.[34] Table 2.2 summarizes the main characteristics and results of each of the reviewed studies.

---

33 For a presentation of the estimation results of several of these studies see Hawdon (1992).

34 The econometric literature on electricity demand in Switzerland consists of three main studies. The first study analyzes residential electricity demand with data at a national level, see Carlevaro and Spierer (1983) and Spierer (1988). The second study examines residential electricity demand using data at a household level, see Dennerlein (1990). However, both studies do not analyze the electricity demand by time-of-use; they aggregate consumption during the peak and off-peak periods. The third study analyses the residential electricity demand by time-of-use using data at a city level, see Filippini (1995b).

Table 2.2 Review of previous studies on the electricity demand by time-of-use

Study	*Data sample*	*Cross-section / Time series/ panel data*	*Model used*	*Model specification**	*Substitutability peak and off-peak consumption*
Hill et al. (1983)	*U.S. household data from rates experiments*	*monthly data for 1976*	*log-linear*	$E_{ih} = f(P_{ih}, y_h, C_j, CDD, HA_{ah})$	*no*
Filippini (1995b)	*Swiss aggregate data at city level*	*panel data for 1987-1990*	*log-linear*	$A_{ij} = f(P_{ij}, Y_j, HDD_j, HS_j, DW, DG)$	*yes*
Caves and Christensen (1980)	*U.S. household data from rates experiment*	*monthly data for 1977*	*translog indirect utility function, CES and generalized Leontieff indirect utility function*	$S_{ph} = f(P_{ph})$	*yes*
Caves et al. (1987)	*U.S. household data from rates experiment*	*monthly data for 1977*	*generalized Leontieff indirect utility function*	$S_{ph} = f(P_{ph}, E_h, HA_{ah}, DC_{ch})$	*yes*
Taylor and Schwarz (1990)	*U.S. household data*	*monthly data for 1985-1988*	*generalized Leontieff indirect utility function*	$S_{mh} = f(P_{ih}, E_h)$	*yes*

* *where* E_{ih} *= electricity consumption by household h in period i, i = peak, mid-peak and off-peak period;* P_{ih} *= electricity price for household h in period i;* y_h *= income of the household h; CDD = cooling degree-days;* HA_{ah} *= electrical appliances of household h, a = 1,2,.., a;* A_{ij} *= per household electricity consumption in city j in period i, i = peak and off-peak period;* P_{ij} *= electricity price in city j in period i;* y_j *= average household income in city j;* HDD_j *= heating degree days in city j;* HS_j *= average household size in city j; DW = dummy variable for the availability a two-part time differentiated tariff to all customers in city h; DG = dummy variable for the natural gas availability in city h;* P_{ph} *= electricity price for household h in period p, p = peak and off-peak period;* E_h *= total electricity expenditure of household h;* S_{ph} *= share of electricity consumption of household h during period p ;* S_{mh} *= share of electricity consumption of household h for electricity commodity m, m = peak electricity, off-peak electricity and maximum demand, and* DC_{ch} *= characteristics of household h, c= 1,2,..., c.*

The first two studies analyze the electricity demand by time-of-use using a system of log-linear demand equations. The log-linear functional form is selected because it is easy to estimate and easy to interpret. However, as presented in the previous section, without imposing severe restrictions on elasticities, it does not adequately reflect the restrictions imposed by the neo-classical theory of consumer behavior. The remaining three studies analyze the allocation of electricity expenditure by using conditional demand system models consistent with neo-classical demand theory and derived from an indirect utility function.

The majority of these studies estimate the electricity demand by time-of-use using data at the household level from rates experiments. In the last few decades in the U.S., several demonstration projects on residential electricity consumption have been promoted in an attempt to better understand the effects of time-of-use pricing on residential electricity consumption. Moreover, time-of-use rate experiments were specifically designed to reduce summer peak demand. Therefore, the data set employed in studies of electricity demand by time-of-use which employ data from rate experiments, concentrate the analysis on summer electricity demand.

Generally, in a rates experiment residential customers of a electric utility were selected randomly and placed on various time-of-use rates for a time horizon that ranges between two and six months. The electric utilities collected monthly data on the electricity usage of each of the selected customers during various daily time periods and thus were able to construct an interesting data set on residential electricity consumption according to time-of-use.

Important considerations on this experimental approach have to be taken into account. The use of data obtained from a rates experiment where consumers were faced with a variety of schedules and prices can create interpretation problems of the estimation results for the following two reasons. First, households in the rate experiment had generally the guarantee that with the experimental bill they would not have to pay more than the usual amount for their electricity consumption. Second, the short time horizon of these rate experiments is not likely to induce a change in the electricity consumption pattern as might be expected from a permanent price adjustment. Moreover, a number of these rate experiments suffer from a lack of price variation in the sample that makes the identification of price and substitution effects difficult.

Hill et al. (1983)

Hill et al. analyze residential electricity demand by time-of-use employing monthly data at the household level from a rate experiment in Arizona, where consumers were faced with a variety of schedules and prices. To promote the voluntary participation to the experiment, the designers introduced an incentive payment. For instance, the customers received the guarantee that they would not have to pay more than the usual amount for their electricity consumption.[35] The interesting feature of the Hill et al. study is that the effects of this incentive payment on the electricity demand by time-of-use have been considered in the model specification.

The data set employed by Hill et al. comprised 80 U.S. households for five months in 1976 and five months in 1977. The model used in this study consists primarily of three log linear equations for peak, mid-peak, and off-peak electricity consumption. Hill et al. presented peak, mid-peak, and off-peak electricity consumption as a linear function of the following variables:

- price during peak period;
- price during mid-peak period;
- price during off-peak period;
- household income;
- cooling degree days;
- total wattage of discretionary-use appliance stocks;
- total wattage of weather-related appliance stocks;
- total wattage of non- discretionary-use appliance stocks;

[35] In a typical rate experiment, each month households received the experimental bill based on time-of-use rates and the bill which would have been incurred under the standard rates (two-part tariff or block tariff). The households paid the bill with the lower amount.

- implicit incentive payment for a household.

The estimated demand functions show an own peak price elasticity of -0.44, an own off-peak period elasticity of -0.23 and a middle period elasticity of -0.47. Moreover, the cross-price elasticities were found to be generally negative and often statistically insignificant.

There are several problems with the Hill et al. model. The log-linear specification, while less restrictive than other specifications, does not account for the restrictions imposed by economic consumer theory. In addition, we believe that the data set utilized by Hill et al. is characterised by the disadvantages presented above concerning data sets that are built on rate experiments. Nevertheless, this study is one of the first innovative attempts to estimate electricity demand by time-of-use as well as to consider effects of incentive payments on elasticity estimates.

Filippini (1995b)

Filippini examined the residential demand for electricity by time-of-use in Switzerland. For this purpose, a model of two equations for peak and off-peak electricity consumption was estimated employing aggregated data referring to four years and 40 cities. Following Hill et al. (1983) Filippini used a double-logarithmic specification. The demand equations for peak and off-peak electricity consumption were estimated by the Ordinary Least Squares (OLS) and by the Error Component (EC) method, which includes a cross-sectional error component. The residential electricity consumption per household during peak periods and the residential electricity consumption per household during off-peak periods were chosen as dependent variables. The explanatory variables were electricity price during a peak period and an off-peak period, household personal income, approximated by the per household collected income taxes by and for municipalities, household size, heating degree days, a dummy variable for the natural gas capabilities of a city and a dummy variable for the cities where all households face a two-part time-differentiated tariff for electricity.

The empirical analysis has highlighted some of the characteristics of the Swiss residential electricity market. The long-run price elasticity was estimated to be -0.71 during the peak period and -1.92 during the off-peak period. These elasticities show a high responsiveness of electricity consumption to changes in prices. Moreover, positive values of the cross-price elasticities show that peak and off-peak electricity are substitutes. The estimated elasticities are higher than those found in all others studies on this subject. However, a comparison of the results is difficult, and typically inconclusive since models, data, and time periods used are not similar. Further, whereas the majority of studies on time-of-day pricing of electricity have their empirical basis in experiments, the study by Filippini is based on observational data that show variation over space and time.

This study has the following two main shortcomings. First, as the Hill et al. study, the employed double-logarithmic specification does not allow to satisfy the theoretical restriction imposed by neoclassical consumer theory. Second, some of the utilities in the sample do not apply time-of-use tariff to all households. Thus, for these utilities the aggregate peak and off-peak usage data represents usage under time-of-use as well as non time-of-use prices. This distorts the usage price relationship. We believe that the introduction of a dummy variable indicating whether the utility applies time-of-use rates to all households, might not be enough to alleviate this problem.

Caves and Christensen(1980)

Caves and Christensen took a different approach to analyzing electricity demand by time-of-use than the two studies outlined above. They estimated a conditional demand system for electricity derived from an indirect utility function. Caves and Christensen, therefore, adopted a two-stage neo-classical model of consumer behavior. Moreover, using extraneous information they made some inferences concerning the total elasticities for the two electricity commodities considered in the analysis, peak and off-peak electricity.

To estimate a conditional demand system for electricity they chose three functional forms, the CES, the translog and the generalized Leontief indirect utility function. Further, the share equations for peak and off-peak electricity consumption were estimated using data at household level from the Wisconsin Pricing Experiment and for the summer months of July and August of 1977. The share of electricity consumption during peak and off-peak periods were regressed against the peak and the off-peak price of electricity.

The estimation results showed that the CES and the generalized Leontief models were more indicative in the analysis of electricity demand by time-of-use. Moreover, the own partial and total price elasticities were estimated to range between -0.22 and -0.81 during the peak period and between -0.19 and -0.85 during the off-peak period. These elasticities showed a moderate responsiveness of electricity consumption to changes in peak and off-peak-prices. Further, the positive values, although small, of the partial elasticity of substitution showed that peak and off-peak electricity are partial substitutes.

While the Caves and Christensen study can be commended for analyzing the electricity demand by time-of-use using a demand system consistent with the neo-classical consumer theory, it has three shortcomings. One of these concerns a possible bias in the estimation results associated with the convenient assumption of homotheticity, which allows the electricity expenditure variable to be excluded from the model, which implies that expenditure shares do not vary with total electricity expenditure levels. The second shortcoming consists in the exclusion from the explanatory variables of important household characteristics such as household size and number and type of electrical appliances. Finally, the disadvantages of estimating an electricity demand system with data from rate experiments presented above are still present in this study.

Caves et al. (1987)

Caves et al. specifically set out to analyze the residential electricity demand by time-of-use using the Hick's aggregation theorem of commodities. Using data on the electricity usage of each customer, which was monitored every fifteen minutes, and applying the Hicksian aggregation theory they were able to model the demand for six different electricity commodities although there were only two pricing periods (peak and off-peak).

Caves et al. estimated a conditional demand system for electricity derived from a generalized Leontief indirect utility function using data of a rate experiment in Wisconsin for the summer months July and August of 1977 and 1978.

In comparison to the previous study of Caves and Christensen (1980), Caves et al. extended the demand system model for electricity in two respects. First, they estimated a homothetic and a non-homothetic version of the electricity demand system. Second, they analyze the influence of household characteristics such as the air-conditioning system, electric water heating ownership and household size on the electricity consumption pattern.

The empirical results show that there were significant substitution possibilities between peak and off-peak electricity consumption. Moreover, using the Hicksian aggregation approach, Caves et al. found that electricity consumption during the critical hours of the system peak was more highly substitutable for off-peak electricity consumption than for consumption during the other peak hours. Furthermore, the estimated expenditure elasticities point to the need for a non-homothetic model specification. Finally, household size and electric water heating ownership were found to influence significantly the electricity consumption pattern.

In conclusion, this study represents an interesting attempt to analyze electricity demand by time-of-use. However, the problems of using data from rate experiments are still present.

Taylor and Schwarz (1990)

Following Caves et al. (1987), Taylor and Schwarz estimated a homothetic and a non-homothetic demand system model for electricity consumption by time-of-use derived from a generalized Leontieff indirect utility function. As in the previous two studies, Taylor and Schwarz employed a two-stage neo-classical model of consumer behavior and considered electricity to be separable from other goods.

The main difference between the study of Taylor and Schwarz and the studies of Hill et al. (1983), Caves and Christensen (1980) and Caves et al. (1987) is that, Taylor and Schwarz utilized observational data on residential electricity consumption instead of data from rates experiments, as did Filippini (1995b). Moreover, the unit of observation was monthly electricity consumption for the individual household broken down into peak and off-peak electricity, and maximum demand within peak hours. Therefore, in addition to peak and off-peak electricity prices, households had to pay a demand charge which was the price for capacity. The latter depended on the maximum level of power consumed in the billing period.

Taylor and Schwarz estimated the share equations for electricity consumption by time-of-use using data of approximately 2000 residential customers of the Duke Power Company for the summer months June, July, and August for the years 1985, 1986, 1987 and 1988.

The empirical results showed a high responsiveness of electricity consumption to changes in electricity prices and demand charge. Moreover, the values of the substitution elasticities indicated that peak electricity consumption and maximum demand are complements while off-peak electricity consumption and maximum demand are substitutes.

This study represents an interesting attempt to analyze electricity demand by time-of-use because a non-experimental data set with a long time horizon is used. However, this study has the shortcoming that household characteristics variables, such as household size and electrical appliances, are not introduced in the model.

After this brief discussion on the results and methods of five studies on electricity demand by time-of-use we would like to suggest several guidelines, which appear to be important in the specification and estimation of a demand system for electricity:

- the estimation of a conditional demand system seems to be a valid approach to the analysis of electricity demand by time-of-use yielding sufficiently precise results for discussing many of the policy implications of time-of-use rates;
- the specification of a conditional demand system derived from an indirect utility function or an expenditure function are more attractive than "ad-hoc" demand systems that work with double-logarithmic specifications, because the former are consistent with neo-classical theory of consumer behavior, while the latter are not;

- the empirical results suggest the need for the estimation of both homothetic and non-homothetic conditional demand system for electricity. Moreover, if the homotheticity assumption is rejected by the data, then , to be consistent with a one-stage procedure of budgeting process, the indirect utility function has to take the generalized Gorman polar form and the overall utility function has to be additive;
- electricity demand models should take into account the influence of household characteristics variables on electricity consumption patterns;
- to identify price and substitution effects the data set used in the empirical analysis should show a high price variation and a long time horizon.

In the next three chapters we present the specification and estimation of the residential electricity demand by time-of-use in Switzerland. Three conditional almost ideal demand system models were estimated using aggregate and micro data sets. Further, the influence of some household characteristics on electricity demand by time-of-use are considered.

CHAPTER 3

REPRESENTATIVE CONSUMER BEHAVIOR MODEL

In section 2.4 of chapter 2 we presented the most important approaches used to identify the parameters of individual consumer behavior from aggregated data. In this chapter, we examine the residential demand for electricity by time-of-use in Switzerland using the representative consumer approach proposed by Muellbauer (1975, 1976).[36]

To this purpose, a conditional almost ideal demand system (AIDS) model for peak and off-peak electricity consumption is estimated employing aggregate panel data at the city level covering the years 1987, 1988, 1989 and 1990.

The chapter is organized as follows. In section 3.1 the representative consumer version of the almost ideal demand system (AIDS) is presented. In section 3.2, the data used in the analysis are discussed, while in sections 3.3 and 3.4, the empirical results and the estimates of the price and substitution elasticities are presented. Some concluding remarks follow in section 3.5.

---

36 This chapter is based on a paper published in the *Energy Journal*. See Filippini (1995a).

3.1 Model specification

Using the representative exact aggregation theory proposed by Muellbauer (1975, 1976), Deaton and Muellbauer (1980a) introduced a version of the AIDS model that satisfies exact aggregation. This version of the AIDS model is derived by generalizing the specific household cost function (2.48) in the following way:

$$ln\, e_h((u_h, \mathbf{p})/k_h) = (1 - u_h)\, ln\, a(\mathbf{p}) + u_h\, ln\, b(\mathbf{p}) \qquad (3.1)$$

where k_h ($h = 1,...,H$) is a household specific parameter that can be interpreted as an index reflecting differences in household size, composition and other household attributes. Using the functional forms for *lna(**p**)* and *lnb(**p**)* presented in section 2.4 and applying Shephard's lemma to the expenditure function (3.1) yields the following budget share equations (See Deaton and Muellbauer, 1980a):

$$w_{ih} = \alpha_i + \sum_j \gamma_{ji}\, ln\, p_j + \beta_i\, ln(\frac{m_h}{k_h P}) \qquad i, j = 1, \ldots\ldots, n. \qquad (3.2)$$

Here w_{ih} is the budget share of the ith good for the hth household, p_j are prices, and $(m_h/k_h\, P)$ is the total real expenditure on all n goods in the consumer's budget, with P a price index defined by

$$ln\, P = \alpha_0 + \sum_j \alpha_j\, ln\, p_j + \frac{1}{2} \sum_i \sum_j \gamma_{ij}\, ln\, p_i\, ln\, p_j$$

and $\gamma_{ij} = (\gamma^*_{ij} + \gamma^*_{ji})/2. \qquad (3.3)$

Application of the representative aggregation theory yields the following aggregate version of (3.2):

$$\overline{w}_i = \alpha_i + \sum_j \gamma_{ji}\ ln\, p_j + \beta_i\ ln(\frac{\overline{m}}{k\,P}) \qquad i,j = 1,\ldots\ldots,n, \tag{3.4}$$

where

$$\overline{w}_i = \frac{\sum_{h=1}^{H} m_h\, w_{ih}}{\sum_{h=1}^{H} m_h} \qquad and \qquad ln(\frac{\overline{m}}{k}) = \frac{\sum_{h=1}^{H} m_h\, ln(\frac{m_h}{k_h})}{\sum_{h=1}^{H} m_h}$$

and $\overline{m}$ is the average level of total expenditure. In empirical applications, households have often been assumed to have common tastes (i.e., $k_h = 1$ for all h), which implies that k becomes an index of the equality of the distribution of households expenditures. Under this assumption, Deaton and Muellbauer (1980a) note that this index conforms to Theil's (1967) entropy measure of equality (Z) deflated by the number of households (H). Therefore, substituting and rearranging, we obtain

$$\overline{w}_i = \alpha_i + \sum_j \gamma_{ji}\, ln\, p_j + \beta_i\, ln(\frac{\overline{m}}{P}) + \beta_i\, ln(\frac{H}{Z}) \qquad i,j = 1,\ldots\ldots,n, \tag{3.5}$$

Assuming that the deflated measure of equality is a constant so that the distribution parameters are absorbed in the intercept, the estimable representative consumer behavior model becomes

$$\overline{w}_i = \alpha_i + \sum_j \gamma_{ji}\, ln\, p_j + \beta_i\, ln(\frac{\overline{m}}{P}) \qquad i,j = 1,\ldots\ldots,n, \tag{3.6}$$

Adding-up, homogeneity, and symmetry can be imposed on this system similar to those equations we have presented in (2.55), (2.56), and

(2.57). Moreover, using Stone's price index (2.54) a "linear approximate almost ideal demand system" can be derived.[37]

The approach used in this study is to represent consumer expenditure allocation problems in stages. Thus, we assume the two-stage budgeting process on the part of consumers presented in section 2.2. In the first stage, the consumer determines the allocation of total expenditure between various commodity groupings, e.g., food, transport, electricity. In the second stage, the demand for time-of-use electricity is determined by the electricity prices during peak and off-peak periods and total electricity expenditures from the first stage. Therefore, the consumer now decides how expenditures on electricity should be allocated between consumption during peak and off-peak periods. The primary objective of of the present study is the estimation of a separable demand system for electricity; therefore, we are dealing with the second stage of the two-stage budgeting process on the part of consumers.[38]

The main issue that can be analyzed with the estimation of a separable demand system for electricity is substitution among electricity commodities, given a household's level of electricity expenditure. The issue of the decrease or increase in total electricity consumption from a change in the overall electricity price index cannot be analyzed because it requires an estimation of the first stage of the budgeting process.

As presented in the previous chapter, a necessary and sufficient condition for the second stage of a two-stage consumption analysis is to assume that consumers possess utility functions which are weakly separable with respect to electricity.[39] Moreover, necessary and sufficient conditions for the budgeting procedure to be consistent with a one-stage procedure are that the group utility functions be homothetic or that the indirect utility functions have the generalized Gorman polar form and the overall utility function is additive. Thus, employing the AIDS model, which illustrates a

---

37 The linear form of the AIDS model presented in equation (3.6) has been widely used in applied demand analysis. For recent applications and discussions of this model see Mdafri and Brorsen (1993) and Alston, Foster and Green (1994).

38 This has been an often cited approach in modelling electricity demand by time-of-use. See Aktinson (1979), Caves and Christensen (1980), Caves, Christensen and Herriges (1987), Taylor and Schwarz (1990) and Mountain and Lawson (1992).

39 See Thomas (1987).

special case of the generalized Gorman polar form, and implicitly assuming an additive overall utility function, our separable demand system for electricity is consistent with a one-stage procedure.

Following Caves et al. (1987) and Taylor and Schwarz (1990) we estimate both homothetic and non-homothetic separable electricity demand models. This allows us to compare the results of both models in term of elasticities.

The AIDS model can be easily modified to account for expenditure shifts determined by differences in household size, structure, taste and climate. Therefore, in this analysis of residential electricity demand by time-of-use, we choose to introduce household size and heating degree days to the AIDS model (3.6).[40] In accordance with procedures set forth by Ray (1980) and Pollack and Wales (1992), these variables are introduced into the model with a log-linear scaling procedure.[41] Thus, starting with equation (3.6) and assuming separability, the non-homothetical linear approximate AIDS representative consumer model for electricity consumption during peak and off-peak periods can be written as

$$w_P = \mu_P + \gamma_{PEP} \ln P_{EP} + \gamma_{PEO} \ln P_{EO} + \beta_P \ln\frac{m}{P} + \eta_P \ln HS + \pi_P \ln HDD \qquad (3.11)$$

$$w_O = \mu_O + \gamma_{OEP} \ln P_{EP} + \gamma_{OEO} \ln P_{EO} + \beta_O \ln\frac{m}{P} + \eta_O \ln HS + \pi_O \ln HDD \qquad (3.12)$$

where w_P = the share of per household electricity expenditure during the peak period in city c (c = 1,..., 21), is obtained by dividing expenditures of peak electricity by total electricity expenditure, P_{EP} = real electricity price during the peak period, P_{EO} = real electricity price during the off-peak

---

40 A dummy variable (DG) was introduced in an earlier study by Filippini (1993) in an attempt to differentiate cities where natural gas is available to households. However, this variable does not have a clear interpretation at the aggregate level. Moreover, a dummy variable (DW) was introduced in order to differentiate cities where part of the weekend is considered as off-peak period of electricity consumption. With a F-test we failed to reject the hypothesis that the regression coefficients of these dummy variables is zero.

41 For an application of the log-linear scaling procedure, see Wellman (1992).

period, (m/P) = per capita total expenditure on electricity divided by the Stone price index for electricity goods (P),[42] HS = household size measured by dividing population by the number of residential households, HDD = heating degree days, and w_O = the share of off-peak expenditure in total electricity expenditure.

The homothetical version of the AIDS model (3.11) and (3.12) is obtained by imposing the electricity expenditure coefficient β_P and β_0 as zero.

3.2 Data and variables

To analyze electricity demand during peak and off-peak periods, panel data covering four years (1987,1988,1989 and 1990) and 21 cities in Switzerland were collected by means of a survey on rates and residential electricity consumption by time-of-use. A mail questionnaire was sent to all 130 urban electric utilities in Switzerland and 67 were returned. Twenty-three questionnaires were returned with incomplete information, resulting in a basic sample of 44 questionnaires which could be used in the analysis. The data for the other variables were taken from the annual publication of the Swiss Cities Association and from the monthly publication of the Swiss Federal Institute of Meteorology.

As stated in chapter 1, the rate schedule facing the majority of Swiss customers is a two-part tariff containing a fixed monthly charge and a constant price per kWh of electricity consumed that varies according to time (day/night). This rate structure is used by all electric utilities in our sample. Further, we can distinguish cities offering this type of tariff to all residential customers and cities that offer it only to customers who use a lot of electricity during off-peak periods, for example for electric space heating. Therefore, for some of the cities in the sample the aggregate peak and off-

42 To avoid simultaneity problems we use the median values of the share to calculate the Stone price index (2.58). In a earlier study by Filippini (1995a) the household individual values of the share were used to calculate the Stone price Index. However, estimating a conditional demand system with only two commodities can give rise to a simultaneity problem.

peak usage data represents usage under time-of-use as well as non time-of-use prices. This could distort usage price relationships. For this reason, we excluded these cities from our data set, resulting in a sample of 21 cities employed in our analysis. Thus, in the budget share equations (3.11) and (3.12) presented in section 3.2, a peak and an off-peak electricity price appears.[43]

Because it was assumed that electricity during peak and off-peak periods is separable from other commodities, the real expenditure variable is given by per capita expenditure for electricity divided by the price index for electricity *(m/P)*.

Assuming that weather might influence household electricity consumption, a climate variable (*HDD* = heating degree days) was included in the model. This variable should account for the impact of weather on the need for space heating. Because the summer temperatures are not extremely high in Switzerland, there is little need for air-conditioning. Thus, the model does not contain cooling degree days as a second climate variable.

Household size (HS) is included in the model in order to account for the impact of the number of members per household on the demand for services provided by electricity-using appliances. A large family is expected to consume more electricity given a certain stock of appliances in the household. Households include single family homes, apartments, duplexes, and condominiums as long as they have their own connection to the network.

With regard to choice of econometric technique, it should be noted that in the econometric literature we can find various types of models focusing on cross-sectional variation, i.e., heterogeneity across units. The three most widely used approaches are: the OLS model, the least squares

43 The fixed fee variable does not enter into the share equations for peak and off-peak electricity consumption. Theoretically, the effect of a fixed fee on electricity consumption should be equal to, but opposite in sign from, the effect of income. In order to impose this constraint, the fixed fee is directly deduced from the real total expenditure or income. For an interesting discussion about the appropriate price to include in an electricity demand equation, see Taylor (1975) and Nordin (1976).

dummy variable (LSDV) model, and the error components model (EC).[44] In the OLS model a common intercept as well as common slope coefficients are assumed across individuals and time periods. The main difference between the OLS and LSDV models involves the constant term. By introducing a complete set of cross-section dummy variables, LSDV allows a separate constant term for each unit. The third type is the error components model. While there are several alternative ways to specify the error components, the most common is to assume that the individual constants are random variables. In this case, differences between units are not viewed as parametric shifts of the regression function as in the LSDV model, but as randomly distributed shocks. EC is often preferred to the use of LSDV since it accomplishes the same objective with a lower cost in terms of degrees of freedom.

Table 3.1 provides some descriptive statistics of the dependent and independent variables.

Table 3.1 Descriptive statistics of dependent and independent variables (1987-1990)

Variables	25th Percentile	Median	75th Percentile
Per household budget share of electricity consumption during peak-period (W_p)	0.69	0.74	0.79
Price during the peak period (P_{EP})	0.16 SwF/kWh	0.17 SwF/kWh	0.19 SwF/kWh
Price during the off-peak period (P_{EO})	0.07 SwF/kWh	0.08 SwF/kWh	0.09 SwF/kWh
Household size (HS)	2.1	2.3	2.4
Per household electricity expenditure per year (m)	338 SwF	433 SwF	512 SwF
Heating degree days (HDD)	3224	3355	3717

44 For a detailed presentation of the econometric methods that have been used to analyse panel data, see Balestra and Nerlove (1966), Hsiao (1986), Greene (1993) and Baltagi and Ray (1992).

3.3 Estimation Results

Because the demand system expressed in budget share form is singular, only one equation (peak electricity demand) is estimated. The parameters of the deleted equation can be calculated residually using the adding-up, homogeneity, and symmetry restriction (2.55), (2.56) and (2.57). The coefficients of the non-homothetical and homothetical demand system equation (3.11) estimated using the OLS, the LSDV and the EC methods are shown in Tables 3.2 and 3.3.[45]

Table 3.2 Non-homothetic residential time-of-use electricity demand (standard errors in parentheses)

	OLS	EC	LSDV
Independent Variables	Dependent variables w_P	Dependent variables w_P	Dependent variables w_P
Constant	1.888*** (0.422)	0.424* (0.220)	0.392 (0.267)
$\ln P_{EP}$	-0.364*** (0.093)	-0.255** (0.110)	-0.182** (0.090)
$\ln P_{EO}$	0.364*** (0.093)	0.255** (0.110)	0.182** (0.090)
$\ln (m/P)$	-0.081* (0.044)	-0.050** (0.021)	-0.061*** (0.028)
ln HS	-0.022 (0.094)	-0.084 (0.105)	-0.076 (0.087)
ln HDD	-0.024 (0.030)	0.019 (0.018)	0.018 (0.018)

*, **, ***: significantly different from zero at the 90%, 95%, 99% confidence level.

45 Tables 3.2 and 3.3 omit the estimated coefficients of the city specific dummy variables; a copy of these coefficients is presented in Appendix A.

The two versions of the AIDS model for electricity consumption by time-of-use rates were estimated with the homogeneity restriction imposed. In the case of only two goods, homogeneity implies that the symmetry and the adding-up properties are automatically satisfied.[46] An F-test was used to test the homogeneity restriction. The F-value shows that homogeneity is an acceptable restriction.

Estimation results of the non-homothetical share equations are satisfactory in so far as the coefficients on electricity prices, which form the basis for the calculation of the elasticities, have the expected sign and are generally significant. Moreover, the significant coefficient of real electricity expenditure (m/P) shows that the budget share of peak and off-peak electricity are sensitive to changes in total electricity expenditure. Therefore, our empirical results suggest the need for a non-homothetic specification.[47] The parameters of household size (HS) and heating degree days (HDD) are not significant. This result indicates that the allocation of electricity to peak and off-peak periods is not influenced by household size or by weather.

Estimation results for the homothetic specification are given in Table 3.3.[48] The estimated price coefficients of the OLS, LSDV and EC models are generally the same as these reported in Table 3.2. Therefore, these results confirm that the shares of electricity demand by time-of-use react on electricity price variations. As in the non-homothetical specification, the parameters of household size (HS) and heating degree days (HDD) are not significant.

Estimation using the LSDV model yielded results which showed the highest standard errors for most of the coefficients, followed by the OLS and the EC models. The price effects are generally smaller in the LSDV model than in the EC or OLS model, whereas the coefficient of real electricity expenditure (m/P) is larger in the OLS model than in the LSDV or EC model.

---

46 While the adding-up restriction is always satisfied by the AIDS model, the symmetry restriction is only in the case of two goods automatically satisfied.

47 See Kohler (1983) for a discussion of the homotheticity assumption in estimating electricity demand by time-of-use.

48 In interpreting these results, it is to remember that by leaving out a relevant explanatory variable like electricity expenditure from a share equation can cause bias in the estimated coefficients of the variables that are in the equation.

Table 3.3 Homothetic residential time-of-use electricity demand (standard errors in parentheses)

	OLS	EC	LSDV
Independent Variables	Dependent variables w_P	Dependent variables w_P	Dependent variables w_P
Constant	1.295***	0.826***	0.896***
	(0.258)	(0.150)	(0.153)
$\ln P_{EP}$	-0.370***	-0.289**	-0.239**
	(0.099)	(0.114)	(0.096)
$\ln P_{EO}$	0.370***	0.289**	0.239**
	(0.099)	(0.114)	(0.096)
ln HS	-0.081	-0.046	-0.030
	(0.105)	(0.108)	(0.102)
ln HDD	-0.026	0.020	0.019
	(0.031)	(0.019)	(0.019)

*, **, ***: significantly different from zero at the 90%, 95%, 99% confidence level.

The point of presenting three similar estimates is that the precision with which the parameter of most interest, the coefficient of the prices, is estimated does differ markedly between the estimation methods. The large value of the Breusch-Pagan LM test-statistic in the OLS model indicates the presence of individual effects.

As noted by Baltagi and Griffin (1984), a relatively small within variation may result in GLS estimates that are very different from LSDV estimates. Because the precision of the LSDV estimator is proportional to the within variation, averaged over different cities, LSDV estimates in cases where the data set contains variables that show a relatively small within variation, will have large standard errors.

According to Baltagi and Griffin (1984), if the between variation for the relevant explanatory variables in the model (i.e., the variation between the cities) greatly exceeds the within variation (i.e., the variation over time), then OLS becomes a preferred estimator. In our model the most important explanatory variables are the electricity prices by time-of-use and the real electricity expenditure, which allow us to compute price and substitution elasticities. In our sample we observed that the between variation of these variables is much larger than the within variation. Therefore, following the arguments of Baltagi and Griffin (1984), the price and substitution elasticities are based on the results of OLS estimates. Moreover, the presented standard errors for the OLS and LSDV models are calculated from (White's) heteroscedastic-consistent variance-covariance matrix.

3.4 Partial elasticity estimates

As mentioned above, the demand system (3.6) was estimated for peak and off-peak electricity consumption by assuming that electricity in the utility function is separable from other goods. The implication of this assumption is that the elasticities calculated with the help of the expressions (2.59), (2.60), and (2.63) from the second stage are partial elasticities. They show the effect of a change in the electricity price during the period j on electricity demand in the period j, holding total electricity expenditure, m, constant. Therefore, elasticity estimates are conditional upon allocation of total expenditure between electricity and other goods. However, these partial elasticities are sufficient to analyze many of the policy impacts of time-of-use tariffs.[49]

Partial own price, cross price, and substitution elasticities for electricity demand by time-of-use, and calculated at the median budget share for both homothetic and non-homothetic model specifications using expressions (2.59), (2.60) and (2.63) are presented in Table 3.4.[50]

---

49 See Caves,Christensen and Herriges (1987).

50 Elasticity estimators in flexible functional forms, such as AIDS and Translog models, are non-linear transformation of the parameter estimators and demand or factor cost shares, so that standard errors for their estimates cannot be calculated exactly. Moreover, this non-linearity makes the usefulness of approximates standard error estimates questionable. Therefore, following several past studies (e.g., Deaton and Muellbauer (1980a); Jorgenson et. al. (1988) and Taylor and Schwarz (1990)), we do

Table 3.4 Estimated partial elasticities

	Non-homothetic model	Homothetic model
Price elasticity, peak	-1.39	-1.48
Price elasticity, off-peak	-2.58	-2.54
Cross price elasticity, peak/off-peak	0.50	0.49
Cross price elasticity, off-peak/peak	1.24	1.54
Elasticity of substitution peak/off-peak	2.97	3.02

The results appear reasonable and in accord with neo-classical consumer theory. The AUES matrix evaluated at the mean data point is negative semi-definite, therefore, the negativity condition of the demand system is satisfied.

The partial own-price elasticities vary from -1.39 to -2.58. These values show that the demand for electricity during both peak and off-peak periods is responsive to changes in prices. A one percent increase in the peak price of electricity will result in about a 1.4 percent decline in the per household consumption of peak electricity, while a 1 percent increase in the off-peak price will result in about a 2.6 percent decline in the per household consumption of off-peak-electricity.

All the cross-price elasticities are positive. These results indicate that electricity consumption can be substituted between the two rate periods to a substantial degree. The positive value of the elasticity of substitution confirms that the goods peak and off-peak electricity are substitutes. Finally, the calculated values of partial expenditure elasticities are 0.89 for peak electricity and 1.33 for off-peak electricity, indicating that off-peak

not report standard errors for elasticity estimators. See Krinsky and Robb (1986) and Eakin et. al. (1990) for a discussion on problems and possible methods to calculate confidence intervals for elasticity estimators in flexible functional forms.

electricity is expenditure elastic and peak electricity is expenditure inelastic. This implies that as electricity expenditure increases, the share of electricity expenditure during the peak period declines, whereas during off-peak period electricity expenditure increases.

The estimated price elasticities are higher than those found in other studies.[51] However, a comparison of the results is difficult, and typically inconclusive since models, data, and time periods used are not similar. Further, the majority of the studies on time-of-day pricing of electricity have their empirical basis in experiments where consumers were faced with a variety of schedules and prices with, however, the guarantee that they would not have to pay more than the usual amount for their electricity consumption.

3.5 Concluding comments

The estimation results of the representative consumer behavior model characterizes the Swiss residential electricity market as rather price responsive. The own partial price elasticities were estimated to range between -1.39 and -1.48 during the peak period and between -2.54 and -2.58 during the off-peak period. These elasticities show a high responsiveness of electricity consumption to changes in peak and off-peak-prices. Moreover, the positive values of the partial cross price elasticities and substitution elasticities show that peak and off-peak electricity are substitutes.

Finally, the results show that household characteristics variables included in the model specification, household size and weather, do not influence the allocation of electricity to peak and off-peak periods. These results are not surprising because aggregate data sets have the disadvantage of containing inaccurate information on household characteristics. In the next chapter, the aggregate panel data set employed in this chapter is combined with micro panel data to obtain more precise estimates of the impact of household characteristics variables on electricity demand by time-of-use.

---

51 See Hawdon (1992) and the second chapter of this book.

CHAPTER 4

AGGREGATE CONSUMER BEHAVIOR MODEL

Generally, the estimation of a demand system for electricity with aggregate data sets allows an accurate estimation of price coefficients, but does not permit a precise estimation of the parameters of the household characteristics variables, because of inadequate information on these variables. To estimate correctly both the coefficients of prices, expenditures, and household characteristics variables, the analyst has to pool information contained in the aggregate and micro data sets available.

Principally, in the literature we find two approaches that combine information on aggregate and micro levels in estimating the coefficients of a demand system. The first approach, suggested by Tobin (1950) and Maddala (1971), consists of estimating the coefficients of some household characteristics variables from micro cross-sectional or panel data, then introducing these estimates into the demand equation, which are estimated with aggregate time-series or panel data, to estimate the price coefficients and coefficients of other variables common in both regressions.[52] The problem with this approach is that the estimated demand functions are not based on the theory of exact aggregation which, as already presented in chapter 2, establishes the conditions under which it is possible to pool micro

---

52 See Hsiao (1986) for a general presentation of the pooling approach suggested by Tobin (1950) and Maddala (1971).

and aggregate data sets to estimate demand system equations. In fact, the suitability and comparability of estimates from aggregate consumer and micro consumer models that are not related by the exact aggregation theory are questionable.

The second approach proposed by Jorgenson, Lau and Stocker (1982) and based on the exact aggregation theory of Lau (1982), consists of simultaneous estimation of prices, expenditure and household characteristics' effects by pooling aggregate data sets with micro data sets.[53]

In this chapter we follow this latter approach because it is, from a theoretical point of view, more sound than the one proposed by Tobin (1950).[54] However, instead of pooling aggregate time series data with micro cross section data like Jorgenson, Lau and Stocker (1982), we pool aggregate panel data with micro panel data. The main advantage in utilizing a pool of panel data instead of a pool of time series and cross section data is that panel data offers the possibility of capturing variations of the variables across different units in space, as well as variations that emerge over time. Moreover, the price and household characteristics variables at the aggregate and micro level reflect the behavior of the units during the same time period. Therefore, in this chapter we present a model of electricity demand by time-of-use that better analyzes the impact of household characteristics variables on electricity consumption than the representative consumer behavior model presented in chapter 3. However, the data requirements needed to estimate the aggregate consumer behavior model by pooling aggregate and micro data sets is rather formidable, thus limiting, the number of household characteristics variables to include in the model.

This chapter is organized as follows: In the first section the aggregate consumer behavior version of the almost ideal demand system (AIDS) is presented. In section 4.2, the data used in the analysis is discussed while in section 4.3 and 4.4, the empirical results and estimates of the elasticities are presented. Some concluding remarks follow in section 4.5.

---

53 See Jorgenson, Slesnick and Stoker (1988) for the estimation of an econometric model of demand for energy based on exact aggregation theory.

54 This chapter is an extension of a research performed jointly with S. Schim Van Der Loeff. I am grateful to my coauthor for his helpful comments and highly competent research collaboration. See Filippini and Schim Van Der Loeff (1994).

4.1 Model specification

Rossi (1988) and Browning (1993) present an implementation of the exact aggregation theory of Lau (1982) to the AIDS model, incorporating household characteristics variables through the translation approach. In order to adapt the AIDS model, presented in micro form in chapter 2, to the aggregate consumer behavior model proposed by Jorgenson, Lau and Stocker (1982), Rossi (1988) generalized the AIDS household specific cost function (2.48) in the following way:

$$ln\, e_h(u_h, p, \mathbf{c}_h) = (1-u_h)\, ln\, a(\mathbf{p}, \mathbf{c}_h) + u_h\, ln\, b(\mathbf{p}, \mathbf{c}_h) \tag{4.1}$$

with

$$\begin{aligned} ln\, a(\mathbf{p}, \mathbf{c}_h) = {} & \alpha_0 + \sum_i \alpha_i\, ln\, p_i + \frac{1}{2}\sum_i \sum_j \gamma_{ij}\, ln\, p_i\, ln\, p_j \\ & + \sum_i \sum_d \eta_{id}\, ln\, p_i\, c_{hd}\,, \end{aligned} \tag{4.2}$$

$$ln\, b(\mathbf{p}, \mathbf{c}_h) = ln\, a(\mathbf{p}, \mathbf{c}_h) + \beta_0 \prod_i p_i^{\beta_i} \tag{4.3}$$

where u_h denotes utility for the hth household whose household characteristics are described by d-vector c_h with elements c_{hd}. The ith commodity price is denoted by p_i. Therefore, the specification (4.1), (4.2) and (4.3) allows the cost function to depend upon prices, utility and household characteristics. The latter are introduced in the model by the translating approach suggested by Jorgenson, Lau and Stocker (1982).

Applying Shephard's lemma to the expenditure function (4.1), one obtains the household AIDS:

$$w_{ih} = \alpha_i + \sum_j \gamma_{ji}\, ln\, p_j + \beta_i\, ln\left(\frac{m_h}{P_h}\right) + \sum_d \eta_{id}\, c_{hd} \tag{4.4}$$

where w_{ih} is the budget share of the ith good for the hth household, p_j are the prices, and (m_h/P_h) is the total real expenditure on all n goods in the consumers budget, with P_h a price index defined by

$$ln\, P_h = \alpha_0 + \sum_j \alpha_j\, ln\, p_j + \frac{1}{2}\sum_i \sum_j \gamma_{ij}\, ln\, p_i\, ln\, p_j + \sum_i \sum_n \eta_{id}\, ln\, p_i\, c_{hd} \qquad (4.5)$$

and $\gamma_{ij} = (\gamma^*_{ij} + \gamma^*_{ji})/2$.

As usual $ln\, P_h$ is approximated by the Stone price index (2.58). To satisfy the properties of demand functions known as adding-up, homogeneity and symmetry, the following restrictions are imposed:

Adding up: $$\sum_i \alpha_i = 1,\ \sum_i \gamma_{ij} = 0,\ \sum_i \beta_i = 0,\ \sum_i \eta_{id} = 0 \qquad (4.6)$$

Homogeneity: $$\sum_j \gamma_{ij} = 0 \qquad (4.7)$$

Symmetry: $$\gamma_{ij} = \gamma_{ji} \qquad (4.8)$$

As shown in chapter 2, the negativity condition cannot be ensured by any restrictions on the parameters alone, but can be checked by evaluating the Allen-Uzawa Elasticities of Substitution (AUES) matrix at some data point.

The empirical analysis of electricity demand by time-of-use will be carried out using household panel data pooled with aggregate panel data. Thus, the question arises as to the conditions under which it will be possible to identify individual parameters from aggregate data. Necessary and sufficient conditions for such identification are stated in the Fundamental Theorem of Exact Aggregation (Lau, 1982) which was introduced in chapter 2 of this book. These conditions state that expenditure shares must be a linear function of the household characteristics c_{hd} , and total electricity expenditure m_h.

The aggregate expenditure share equations are given by

$$w_i = \frac{\sum_{h=1}^{H} m_h w_{ih}}{\sum_{h=1}^{H} m_h} \tag{4.9}$$

According to the individual share equations (4.4), and following Rossi (1988) and Browning (1993), the aggregate expenditure shares can be expressed as

$$w_i = \frac{\sum_{h=I}^{H} m_h w_{ih}}{\sum_{h=I}^{H} m_h} = \alpha_i + \sum_j \gamma_{ji} \ln p_j + \beta_i \left(\sum_h \frac{m_h \ln m_h}{\sum_h m_h} \right) - \beta_i \ln P$$

$$+ \sum_d \eta_{id} \left(\frac{\sum_h m_h c_{hd}}{\sum_h m_h} \right) = \alpha_i + \sum_{j=I}^{n} \gamma_{ji} \ln p_j + \beta_i \ln \left(\frac{\sum_h m_h}{P} \right) \tag{4.10}$$

$$+ \beta_i \left(\sum_h \frac{m_h}{\sum_h m_h} \frac{\ln m_h}{\sum_h m_h} \right) + \sum_d \eta_{id} \left(\frac{\sum_h m_h c_{hd}}{\sum_h m_h} \right)$$

This form of budget shares employed by Rossi (1988) and Browning (1993) allows estimation of the demand system from the aggregate data. The budget shares depend on the prices of commodities, p_j, on the total real expenditure, $\ln\left(\sum_h m_h / P\right)$, the negative of Theil's inequality measure (see Theil 1967), $\sum_h \frac{m_h}{\sum_h m_h} \frac{\ln m_h}{\sum_h m_h}$, and summary statistics reflecting the joint distribution of characteristics and expenditures across households, $\left(\sum_h m_h c_{hd} / \sum_h m_h\right)$. If household characteristics can be represented as dummy variables, equal to 1 for a household having that characteristic and 0 otherwise, these summary statistics are equal to the share of the household group with characteristic c in aggregate total expenditure.

The household budget shares (4.4) and the aggregate budget shares (4.10) permit the estimation of a conditional demand system for electricity using household panel data pooled with aggregate panel data.[55]

Assuming that consumers possess utility functions which are weakly separable with respect to electricity and follow a two-stage budgeting process described in the previous chapters, the household budget shares of peak and off-peak electricity consumption can be expressed as a function of electricity prices, real electricity expenditures and household characteristics. The aggregate budget shares of peak and off-peak electricity consumption depend on electricity prices, on aggregate total real electricity expenditure corrected by the Theil inequality measure, and on the share of the household group with characteristic c in aggregate total electricity expenditure.

Besides household characteristics, stock of electrical appliances and household size, we introduce weather as an additional variable. Since the majority of Swiss urban households live in rented apartments, they cannot readily choose whether to use certain electrical appliances, among which some of them most energy intensive such as an electric cooker or water heater. Therefore, the stock of electrical appliances is considered an exogenous household characteristic in the following.

---

55 It is clear that the household and the aggregate budget shares (4.4) and (4.10) can also be estimated separately using micro and aggregate data sets, respectively. However, in cases where the micro or/and aggregate data sets do not contain information on all explanatory variables the simultaneous estimation becomes an interesting approach to obtain more precise estimation results.

With regard to household size, we distinguish one-person households, households with two or three persons, and households with four or more persons. With regard to stocks, households having a small or large stock of electrical appliances are distinguished.[56]

Thus, starting with equations (4.4) and assuming separability, the household AIDS consumer model for electricity consumption during the peak and off-peak periods can be written as

$$\begin{aligned} w_{hP} = {} & \mu_{hP} + \gamma_{PEP} lnP_{EP} + \gamma_{PEO} lnP_{EO} + \beta_P ln\frac{m_h}{P} + \eta_{PHS1} D_{HS1} + \eta_{PHS2} D_{HS2} \\ & + \eta_{PHS3} D_{HS3} + \eta_{PS1} D_{S1} + \eta_{PS2} D_{S2} + \eta_{HDD} D_{HDD} \end{aligned} \quad (4.11)$$

$$\begin{aligned} w_{hO} = {} & \mu_{hP} + \gamma_{OEP} lnP_{EP} + \gamma_{OEO} lnP_{EO} + \beta_O ln\frac{m_h}{P} + \eta_{OHS1} D_{HS1} + \eta_{OHS2} D_{HS2} \\ & + \eta_{OHS3} D_{HS3} + \eta_{OS1} D_{S1} + \eta_{OS2} D_{S2} + \eta_{HDD} D_{HDD} \end{aligned} \quad (4.12)$$

where w_{hP} = per household share of electricity expenditure during the peak period, obtained by dividing expenditures on peak electricity by total electricity expenditure, w_{hO} = per household share of electricity expenditure during the off-peak period, P_{EP} = electricity price during the peak period, P_{EO} = electricity price during the off-peak period, *(m/P)* = per household total expenditure on electricity divided by the Stone price index for electricity goods (*P*), D_{HS1} = dummy variable indicating whether the household is a one-person household, D_{HS2} = dummy variable indicating whether the household is a two or three persons household, D_{HS3} = dummy variable indicating whether the household is a four or more household, D_{S1} = dummy variable indicating whether the household has a small stock of electrical appliances, D_{S2} = dummy variable indicating whether the household has a large stock of electrical appliances and D_{HDD} = dummy

56 Households are distinguished only with regard to their stock of appliances because no additional household characteristics are available at the aggregate level.

variable indicating whether the household is living in a city characterized by more than 3,300 heating degree days a year.[57]

The aggregate AIDS consumer model for electricity consumption during the peak and off-peak period can be written as

$$w_{Pc} = \mu_P + \gamma_{PEP}\, lnP_{EP} + \gamma_{PEO}\, lnP_{EO} + \beta_P(ln\frac{m}{P} + TH) + \eta_{PHS1}\, S_{HS1} + \eta_{PHS2}\, S_{HS2} + \eta_{PHS3}\, S_{HS3} + \eta_{PS1}\, S_{S1} + \eta_{PS2}\, S_{S2} + \eta_{HDD}\, D_{HDD} \quad (4.13)$$

$$w_{Oc} = \mu_O + \gamma_{OEP}\, lnP_{EP} + \gamma_{OEO}\, lnP_{EO} + \beta_O(ln\frac{m}{P} + TH) + \eta_{OHS1}\, S_{HS1} + \eta_{OHS2}\, S_{HS2} + \eta_{OHS3}\, S_{HS3} + \eta_{OS1}\, S_{S1} + \eta_{OS2}\, S_{S2} + \eta_{HDD}\, D_{HDD} \quad (4.14)$$

where w_{Pc} = aggregate share of electricity expenditure during the peak period in city c (c= 1,..., 21), w_{Oc} = aggregate share of electricity expenditure during the off-peak period in city c, P_{EP} = real electricity price during the peak period, P_{EO} = real electricity price during the off-peak period, (m/P) = aggregate total expenditure on electricity divided by the Stone price index for electricity goods (P) and corrected by the negative of the Theil inequality index (TH), S_{HS1} = share of the one-person household group in aggregate total electricity expenditure, S_{HS2} = share of the two or three person household group in aggregate total electricity expenditure, S_{HS3} = share of the more than three person household group in aggregate total electricity expenditure, S_{S1} = share of the small stock of electrical appliances household group in aggregate total electricity expenditure, S_{S2} = share of the large stock of electrical appliances household group in aggregate total electricity expenditure and D_{HDD} = dummy variable indicating whether the city is characterized by less than 3300 heating degree days a year.

---

57 Households living in the same city or state are faced with the same heating degree days. Therefore, employing micro data sets refer to households living in the same city does not allow empirical identification of the parameter on this wheather variable.

4.2 Data and variables

As described before, aggregate panel data and household panel data are combined to

obtain more precise estimates of the effects of variations in electricity prices and household characteristics on electricity demand by time-of-use. The aggregate panel data, pertaining to Swiss cities, contains substantial variation in electricity prices, whereas the micro data set, regarding households in the city of Zurich, does not show electricity price variation. Therefore, the estimation of price elasticities with this micro data set is impossible.

The (unbalanced) panel household data set utilized contains data covering the years 1989-1990 on household electricity consumption and household characteristics with a total of 280 observations of households in Zurich. These data were collected in the first instance by a private marketing company IHA (Institut für Marktanalysen AG), that periodically performs a stratified consumer demand survey on about 1,600 households living in Swiss cities. The households are asked to provide information on household demographic characteristics and household electrical appliances. While providing a wide coverage of demographic and electrical appliance variables on each specific household, this consumer demand survey does not contain information on electricity prices and electricity consumption. Therefore, we complement this data set with information from the electric utility of Zurich on the electricity consumption of each household in the survey living in Zurich. Of the about 160 households living in Zurich contained in the sample of IHA for 1989 and 1990 , 5% did not grant us permission to ask the electric utilities about their electricity consumption.

The household characteristics variables assume a value of 1 or 0, depending on whether a household does or does not exhibit a given characteristic.

The stock of electrical appliances is measured using the following electrical appliances index (*EAI*)

$$EAI_c = \sum_{a=1}^{A} CAP_a , \qquad a = 1,2,\ldots,A. \qquad (4.15)$$

where CAP_a is the capacity of electrical appliance *a* expressed in Watts. A household has a small stock of electrical appliances if *EAI* is lower than 3,500 Watts. In sum, this data set is rich in information on household characteristics but does not contain price variation.

The aggregate panel data consists of observations on aggregate electricity expenditure shares, electricity expenditures by demographic groups and electricity prices covering four years (1987, 1988, 1989, 1990) and 21 cities in Switzerland. As already presented in the previous chapter, this information was collected by means of a survey on rates and residential electricity consumption by time-of-use. A questionnaire was sent to all 130 urban electric utilities in Switzerland, of which 67 were returned. Among these 67 cities only 21 electric utilities offer time differentiated tariffs to all residential customers.

For each city we need to know the electricity consumption of one-person households and that of households with four or more persons as a share in total electricity consumption of that city. Similarly, we need to know the share in a city's total electricity bill of households having an EAI in excess of 3,500 Watts. Unfortunately these variables are not directly available.

To construct these variables we first estimate the distribution of households by size and by capacity for each city of the sample.

In the case of Zurich (the city for which we have detailed information at the household level) the distribution of households by size resembles a Poisson distribution which is a one-parameter function that adequately describes a large number of real-world phenomena.[58] The Poisson distribution function $p(x, \lambda)$ is defined as

$$p(x,\lambda)=\frac{e^{-\lambda}\,\lambda^{x}}{x!} \qquad \textit{for } x = 0, 1, \ldots.. \qquad (4.16)$$

58 For a detailed presentation of the Poisson distribution see Larsen and Marx (1986).

where x is a random variable with $E(x) = \lambda$ and $Var(x) = \lambda$. Therefore, if one knows λ then one can construct the distribution of the Poisson random variable x.

Under the assumption that for all Swiss cities the household size is a Poisson random variable, the frequencies of different household sizes can be constructed for each city using the average household size as an estimate of the parameter λ of the Poisson distribution function. Information on the average household size by city can be obtained from *Statistik der Schweizer Städte*.

To obtain the distribution of capacity classes for each city of the sample we again refer to the information available from the sample of Zurich households. From this sample the joint distribution of income I^Z_i (i = 1, 2, 3, 4) and capacity C^Z_w (w = small (s), large(l)) classes can be constructed. Table 4.1 presents how we constructed this joint distribution in the case of Zurich households.

For the other Swiss cities considered in the analysis we obtained information on the distribution among income classes from the statistics on the tax-payers income classes reported in *Statistik der Schweizer Städte*. Therefore, for each city (c) it was possible to calculate the marginal probability of the income classes $p(I^c_i)$.

Table 4.1 Construction of the joint probability distribution of income (I^Z_i) and capacity (C^Z_w), $p(C^Z_w, I^Z_i)$ and the univariate probability distribution $p(I^Z_i)$, $p(C^Z_w)$ or the sample of Zurich households

	C^Z_s	C^Z_l	*Row total*
I^Z_1	$p(C^Z_s, I^Z_1)$	$p(C^Z_l, I^Z_1)$	$p(I^Z_1)$
I^Z_2	$p(C^Z_s, I^Z_2)$	$p(C^Z_l, I^Z_2)$	$p(I^Z_2)$
I^Z_3	$p(C^Z_s, I^Z_3)$	$p(C^Z_l, I^Z_3)$	$p(I^Z_3)$
I^Z_4	$p(C^Z_s, I^Z_4)$	$p(C^Z_l, I^Z_4)$	$p(I^Z_4)$
Column total	$p(C^Z_s)$	$p(C^Z_l)$	

Under the assumption that

$$\frac{p(C_w^Z, I_i^Z)}{p(I_i^Z)} = \frac{p(C_w^c, I_i^c)}{p(I_i^c)} \tag{4.17}$$

i.e., the conditional distribution of capacity, given income, is equal across Swiss cities, the marginal capacity distribution $p\ (C^c{}_w)$ in each Swiss city can be calculated residually.

Table 4.2 presents how the joint distribution of income ($I^c{}_i$) and capacity ($C^c{}_w$), $p\ (C^c{}_w, I^c{}_i)$ for each city (c) of the sample has been derived.

Table 4.2 Construction of the joint probability distribution of income ($I^c{}_i$) and capacity ($C^c{}_w$), p ($C^c{}_w$, $I^c{}_i$) and the univariate probability distribution p ($I^c{}_i$), p ($C^c{}_w$) for each city (c) of the sample

	$C^c{}_s$	$C^c{}_l$	*Row total*
$I^c{}_1$	$\frac{p(C_s^Z, I_1^Z)p(I_1^c)}{p(I_1^Z)} = p(C_s^c, I_1^c)$	$\frac{p(C_l^Z, I_1^Z)p(I_1^c)}{p(I_1^Z)} = p(C_l^c, I_1^c)$	$p\ (I^c{}_1)$
$I^c{}_2$	$\frac{p(C_s^Z, I_2^Z)p(I_2^c)}{p(I_2^Z)} = p(C_s^c, I_2^c)$	$\frac{p(C_l^Z, I_2^Z)p(I_2^c)}{p(I_2^Z)} = p(C_l^c, I_2^c)$	$p\ (I^c{}_2)$
$I^c{}_3$	$\frac{p(C_s^Z, I_3^Z)p(I_3^c)}{p(I_3^Z)} = p(C_s^c, I_3^c)$	$\frac{p(C_l^Z, I_3^Z)p(I_3^c)}{p(I_3^Z)} = p(C_l^c, I_3^c)$	$p\ (I^c{}_3)$
$I^c{}_4$	$\frac{p(C_s^Z, I_4^Z)p(I_4^c)}{p(I_4^Z)} = p(C_s^c, I_4^c)$	$\frac{p(C_l^Z, I_4^Z)p(I_4^c)}{p(I_4^Z)} = p(C_l^c, I_4^c)$	$p\ (I^c{}_4)$
Column total	$p\ (C^c{}_s)$	$p\ (C^c{}_l)$	

To obtain the share of each household group in the city's electricity bill we made the further assumption that the ratios of the average expenditures of any two capacity and household size classes correspond to those found in the Zurich sample. For example, in the case of the two capacity classes this assumption can be expressed as

$$\frac{E_s^Z}{n_s^Z} / \frac{E_l^Z}{n_l^Z} = \frac{E_s^c}{n_s^c} / \frac{E_l^c}{n_l^c} \tag{4.18}$$

where E^Z_s is the total electricity consumption of capacity class s in Zurich, E^Z_l is the total electricity consumption of capacity class l in Zurich, E^C_s is the total electricity consumption of capacity class s in city c ($c = 1, 2, ...21$), E^C_l is the total electricity consumption of capacity class l in city c, n^Z_s is the number of households of capacity class s in Zurich, n^Z_l is the number of households of capacity class l in Zurich, n^C_s is the number of households of capacity class s in city c and n^C_l is the number of households of capacity class l in city c.

It can be easily shown that

$$\frac{E_s^Z}{n_s^Z} / \frac{E_l^Z}{n_l^Z} = \frac{E_s^Z}{p(C_s^Z)} / \frac{E_l^Z}{p(C_l^Z)} = \frac{S_s^Z}{S_l^Z} / \frac{p(C_s^Z)}{p(C_l^Z)} \tag{4.19}$$

where S^Z_s and S^Z_l are the shares of capacity class s and l in the electricity bill of Zurich. Therefore,

$$\frac{S_s^Z}{S_l^Z} / \frac{p(C_s^Z)}{p(C_l^Z)} = \frac{S_s^C}{1-S_s^C} / \frac{p(C_s^C)}{p(C_l^C)} \tag{4.20}$$

or

$$\frac{S_s^Z}{S_l^Z} / \frac{p(C_s^Z)}{p(C_l^Z)} = \frac{S_s^C}{1-S_s^C} / \frac{p(C_s^C)}{p(C_l^C)} \tag{4.21}$$

Knowing S^Z_s and S^Z_l from the sample of Zurich households and knowing $p\,(C^Z_s)$, $p\,(C^Z_s)$, $p\,(C^C_s)$ and $p\,(C^C_s)$ derived using Table 4.1 and 4.2 it is easy to calculate the share of capacity class s in the total electricity bill of city c (S^c_s). Because the information on $p(I^c_j)$ is available for only the year 1987/1988, the variables S^c_s and S^c_l are time-invariant. Following the same method, the share of households with one person and that of households with four and more persons in the total electricity bill of each city c has been calculated.

Tables 4.3 and 4.4 present details on the variables and their values.

Table 4.3 Descriptions of sample household characteristics for the sample of households living in Zurich in 1989-1990

Household characteristics	Variable
Price during the peak period (PEP)	0.15 SwF/kWh
Price during the off-peak period (PEO)	0.05 SwF/kWh
Frequency (%) of small-sized households	25
Frequency (%) of large-sized households	20
Frequency (%) of large capacity households	23

Table 4.4 Description of the variables

Aggregate share equations variables	25th Percentile	Median	75th Percentile
Budget share of electricity consumption during peak-period (W_p)	0.69	0.74	0.79
Price during the peak period (P_{EP})	0.16	0.17	0.19
Price during the off-peak period (P_{EO}) [in SwF/kWh]	0.07	0.08	0.09
Share of small households in aggregate total electricity expenditure (S_{HS1})	0.12	0.13	0.15
Share of the large household group in aggregate total electricity expenditure (S_{HS3})	0.30	0.34	0.37
Share of households having large capacity in aggregate total electricity expenditure (S_{S2})	0.32	0.33	0.33
Total electricity expenditure corrected with the negative of the Theil entropy measure (m) [in SwF]	2,350,650	3,430,040	4,107,600
Heating degree days (HDD)	3224	3355	3717
micro share equations variables	**25th Percentile**	**Median**	**75th Percentile**
Budget share of electricity consumption during peak-period (W_p)	0.85	0.88	0.92
Per household total electricity expenditure (m) [in SwF]	217	308	480

4.3 Estimation results

As in Jorgenson, Stoker and Slesnick (1982), we shall combine the two sets of data, one consisting of observations on electricity expenditures of individual households in Zurich for the years 1989 and 1990, and the other for observations on aggregate electricity expenditure of 21 cities in Switzerland during the four years 1987, 1988, 1989, and 1990. The variables of the (unbalanced) household panel show virtually no variation over time which is why it will be treated as two cross-sections giving rise to separate intercepts for those two years, and in estimating the aggregate city data, form a panel, calling for a choice between the OLS, LSDV and EC estimators presented in the previous chapter. Since by employing the LSDV estimator it is not possible to estimate the effect on electricity consumption of important but non-time varying regressors such as the share of capacity class s in the total electricity bill of city c ($S^c{}_s$), we estimate the OLS and EC models.

As in the case of the representative consumer model presented in chapter 3, we estimate both non-homothetic and homothetic models. Moreover, the homogeneity restriction was imposed, which in the case of only two goods implies that the symmetry and the adding-up properties are automatically satisfied.[59]

For the N = 21 cities in each of the T = 4 years 1987, 1988, 1989, 1990 we denoted:

y the NT vector of observations on day-time electricity expenditure as a share of total expenditure;

X the NT x 2 matrix containing the observations on the logarithm of the ratio of peak to off-peak prices and the dummy variable for heating degree days.

---

59 Since the demand systems expressed in budget share form are singular, only the equations on peak electricity demand are jointly estimated. The parameters of the deleted equations can be calculated residually.

Z the NT x 4 matrix containing the observations on the logarithm of the total electricity expenditure corrected with the Theil inequality measure, on expenditures for electricity according to various socio-economic categories, i.e., small households, large households and households with a large capacity of appliances as a share of expenditure on electricity by all households of a given city.

The estimating equation for the city sample can be written as

$$\mathbf{y} = \alpha_0(\iota_N \otimes \iota_T) + \mathbf{X}\beta + \mathbf{Z}\gamma + \mathbf{u} \tag{4.22}$$

where α_0 is an unknown scalar, β , and γ are vectors of unknown coefficients and ι_N and ι_T are the vectors of length N and T, respectively, with all elements equal to unity. Different assumptions will be made about V($\mathbf{u}$), the variance-covariance matrix of $\mathbf{u}$:

$$(\mathbf{u}) = \sigma^2 \mathbf{I}_{NT} \qquad \text{(OLS)} \tag{4.23a}$$

$$(\mathbf{u}) = \mathbf{I}_N \otimes \sigma^2 \mathbf{I}_T + \sigma_\alpha^2 \mathbf{I}_N \otimes \iota_T \iota_T' \qquad \text{(EC)} \tag{4.23b}$$

Turning to the estimation equation of the household sample, define

v_{it} = the share day-time electricity expenditure in total electricity expenditure;

w_{it} = the 4 x 1 vector of variables containing;

$w_{it,exp}$ = total electricity expenditure;

$w_{it,1}$ = 1 if a household consists of 1 person, 0 otherwise;

$w_{it,2}$ = 1 if a household consists of 4 persons or more, 0 otherwise;

$w_{it,3}$ = 1 if a household has installed capacity in excess of 3,500 Watts, 0 otherwise.

Then, the estimating equations for the household sample are

$$\mathbf{v}_{it} = \alpha_t + \mathbf{w}_{it}^{'}\gamma + \varepsilon_{it}$$
$$i = 1,2,\ldots, M_t; \quad t = 1,2 \quad and \quad \varepsilon_{it} \quad is \ i.i.d \ (0, \sigma_t^2) \tag{4.24}$$

Ignoring the correlation between the cross-sections and the aggregate observations pertaining to Zurich, the joint estimation of the unknown coefficients in (4.22) and (4.24) is carried out by minimizing

$$S(\alpha_1,\alpha_2,\beta,\gamma) = (\mathbf{y}-\mathbf{X}\beta-\mathbf{Z}\gamma)'\mathbf{Q}(\mathbf{y}-\mathbf{X}\beta-\mathbf{Z}\gamma) + \sum_{i,t}(v_{it}-\alpha_t-\mathbf{w}_{i,t}^{'}\gamma)^2/\sigma_{tt} +$$
$$\frac{\rho^2}{1-\rho^2}\left[\sum_t\sum_{i\in I_C}\left(v_{it}-\alpha_t-\mathbf{w}_{i,t}^{'}\gamma\right)^2/\sigma_{tt} + 2\sum_{i\in I_C}\left(v_{it}-\alpha_1-\mathbf{w}_{i,1}^{'}\gamma\right)\left(v_{it}-\alpha_2-\mathbf{w}_{i,2}^{'}\gamma\right)/\sigma_{12}\right] \tag{4.25}$$

where σ^2 and σ_{ts} (t,s = 1,2) are estimated in a first round using, respectively, only the sample of observations on aggregate variables and the observations on individual households.[60] The second line of (4.25) results from taking into account the correlation between the observations i I_C on those households that are in the sample in both years. σ_{12} is the covariance between the observations on these households and ρ is the resulting correlation coefficient. **Q** is an NT x NT transformation matrix that depends on the estimation method employed and concerns itself with the intercept α_0 . **Q** may, in the case of EC, depend on parameters estimated in a previous round.

The estimation results for the non-homothetic model are presented in Table 4.5. The estimates are satisfactory in the sense that they agree with our a priori ideas about size and sign of the coefficients. The coefficient of electricity price has the expected sign and is significant in both the OLS and

---

60 For a presentation of the econometrics of pooling aggregate and disaggregate data see Hsiao (1986) and Jorgenson et. al. (1982).

EC models. The coefficient on real electricity expenditure is significantly different from zero in both models. This result confirms that homotheticity is not a valid assumption in estimating electricity demand by time-of-use. The share of day-time electricity use of small (one-person) and large (four or more persons) households are not significantly different from zero in both models. As expected, households with larger capacity (extensively the effect of the presence in the household of a warm-water boiler) do substitute off-peak for peak electricity consumption.

The point of presenting three similar estimates is analogous to the consideration described in the previous chapter, which argues that the precision with which the parameter of most interest, i.e. the coefficients of the price variables are estimated, differs markedly between estimation methods. The large value of the Breusch-Pagan LM test-statistic in the first-round OLS regression using aggregate (panel-) data pertaining to the 21 cities, indicates the presence of individual effects in spite of the fact that these have been proxied by the inclusion of the measures described above to obtain aggregate household characteristics variables.

Although not significantly different, the absolute values of the EC estimate of the price coefficients are lower than those of the OLS estimates. As in chapter three, we follow the suggestion of Baltagi and Griffin (1984) which proposes resorting to the OLS estimator when the relatively small within variation renders the EC (as well as the within and difference) estimator impractical.

Given the high intertemporal correlation of the aggregate observations, the standard errors will be underestimated by the OLS estimated standard errors (and heteroscedasticity robust standard errors). Therefore, we present standard errors that reflect a lack of information in particular with respect to the price coefficient pertaining to the aggregate sample.

Table 4.5 Joint estimates of the non-homothetical time-of-use share equation (standard errors in parentheses)

Independent Variables	OLS [a] Dependent variable wp	OLS (&) [a] Dependent variable wp	EC[b] Dependent variable wp
$\ln P_{EP}$	-0.410***	-0.410**	-0.312***
	(0.076)	(0.176)	(0.101)
$\ln P_{EO}$	0.410***	0.410**	0.312***
	(0.076)	(0.176)	(0.123)
$\ln (m/P)$	0.035***	0.035**	0.028***
	(0.007)	(0.014)	(0.009)
D_{HDD}	0.004	0.004	-0.001
	(0.016)	(0.022)	(0.002)
S_{HS1}	0.005	0.005	0.001
	(0.014)	(0.016)	(0.014)
S_{HS3}	0.018	0.018	0.020
	(0.015)	(0.016)	(0.015)
S_{S2}	-0.119***	-0.119***	-0.116***
	(0.013)	(0.014)	(0.013)

*, **, ***: *significantly different from zero at the 90%, 95%, 99% confidence level.*

(&) the standard errors in this column have been obtained under the assumption that the intertemporal covariance matrix in the aggregate sample is equal to Σ rather than $\sigma^2 \mathbf{I}_T$.

a. $\mathbf{Q} = \sigma^{-2} \left(\mathbf{I}_{NT} - \frac{\iota_N \iota_N'}{N} \otimes \frac{\iota_T \iota_T'}{T} \right)$ *where σ^2 has been estimated by means of a first round regression, while σ_{ts} $(t, s = 1, 2)$ are OLS estimates from the first round.*

b. $\mathbf{Q} = \sigma^{-2} \left[\left(\mathbf{I}_N \otimes \mathbf{E}_T \right) + \theta\, \mathbf{Q}_B \right]$ *where $\theta = \sigma^2 / \left(\sigma^2 + T \sigma_\alpha^2 \right)$ with σ^2 and σ_α^2 the variance components computed in a first round according to Wallace and Hussain (1969), while σ_{ts} $(t, s = 1, 2)$ are OLS estimates from the first round.*

Table 4.6 Joint estimates of the homothetical time-of-use share equation (standard errors in parentheses)

Independent Variables	OLS [a] Dependent variable wp	OLS (&) [a] Dependent variable wp	EC[b] Dependent variable wp
$\ln P_{EP}$	-0.404***	-0.404**	-0.317***
	(0.087)	(0.197)	(0.113)
$\ln P_{EO}$	0.404***	0.404**	0.317***
	(0.087)	(0.197)	(0.113)
D_{HDD}	-0.013	-0.013	-0.001
	(0.018)	(0.025)	(0.002)
S_{HS1}	-0.015	-0.015	-0.015
	(0.013)	(0.013)	(0.013)
S_{HS3}	0.029*	0.029*	0.030**
	(0.015)	(0.015)	(0.015)
S_{S2}	-0.107***	-0.107***	-0.107***
	(0.013)	(0.013)	(0.013)

*, **, ***: *significantly different from zero at the 90%, 95%, 99% confidence level.*

(&) the standard errors in this column have been obtained under the assumption that the intertemporal covariance matrix in the aggregate sample is equal to Σ *rather than* $\sigma^2 \mathbf{I}_T$.

a. $\mathbf{Q} = \sigma^{-2} \left(\mathbf{I}_{NT} - \frac{\iota_N \iota_N'}{N} \otimes \frac{\iota_T \iota_T'}{T} \right)$ *where* σ^2 *has been estimated by means of a first round regression, while* σ_{ts} $(t,s = 1,2)$ *are OLS estimates from the first round.*

b. $\mathbf{Q} = \sigma^{-2}\left[\left(\mathbf{I}_N \otimes \mathbf{E}_T\right) + \theta\, \mathbf{Q}_B\right]$ *where* $\theta = \sigma^2 / \left(\sigma^2 + T\sigma_\alpha^2\right)$ *with* σ^2 *and* σ_α^2 *the variance components computed in a first round according to Wallace and Hussain (1969), while* σ_{ts} $(t,s = 1,2)$ *are OLS estimates from the first round.*

Table 4.6 presents the estimation results for the homothetical model. The values of the price coefficients are generally the same as these found in the non-homothetical model. The share of day-time electricity use of small (one-person) households is not significantly smaller, while the share of day-time electricity use of large (four or more persons) households is significantly larger.

As in the non-homothetical model, the households with larger capacity do substitute off-peak for peak electricity consumption.

4.4 Partial elasticity estimates

Table 4.4 presents the partial own price, cross price, and substitution elasticities for electricity demand by time-of-use,calculated at the mean budget share, for both homothetic and non-homothetic aggregate consumer model specification.

Table 4.7 Estimated partial elasticities

	Non-homothetic model	Homothetic model
Price elasticity, peak	-1.58	-1.54
Price elasticity, off-peak	-2.54	-2.53
Cross price elasticity, peak/off-peak	0.54	0.54
Cross price elasticity, off-peak/peak	1.67	1.53
Elasticity of substitution peak/off-peak	3.13	3.07

The estimated price and substitution elasticities calculated with the estimation results of the aggregate consumer behavior model confirm the results obtained by the estimates of the representative consumer behavior model, i.e., they have the same sign and similar values.The negativity condition of the demand system is satisfied and, the results are therefore, in accord with neo-classical consumer theory. The values of the partial own-price elasticities show that the demand for electricity during both peak and off-peak periods is responsive to changes in prices. The positive values of cross-price elasticities and of the elasticity of substitution indicate that electricity consumption can be substituted between the two rate periods to a substantial degree. Finally, the estimated partial expenditure elasticities are 1.05 for peak electricity and 0.87 for off-peak electricity. These expenditure elasticities do differ substantially from those found in the previous chapter by estimating the representative consumer behavior model. They indicate that peak electricity is expenditure elastic (rather than inelastic) and off-peak electricity is expenditure inelastic (rather than elastic). This dissimilarity in estimated expenditure elasticities can also be found by comparing the studies by Caves et. al. (1987) and by Taylor and Schwarz (1990). The explanation underlying this difference may be determined by different model specifications and data sets employed in this chapter and in chapter 3 as well as in studies by Caves et. al. (1987) and Taylor and Schwarz (1990).

4.5 Concluding comments

The estimation results of the aggregate consumer behavior model are in line with the estimation results of the representative consumer behavior model, which characterizes the Swiss residential electricity market as rather price responsive. The partial own price elasticities were estimated to a range between -1.54 and -1.58 during the peak period and between -2.53 and -2.54 during the off-peak period. Moreover, the positive values of the partial cross price elasticities and substitution elasticities show that peak and off-peak electricity are substitutes. In addition, the empirical results reject the homotheticity assumption of the model. However, in contrast with conclusions drawn from the estimation of the representative consumer behavior model presented in the previous chapter, peak electricity is expenditure elastic and off-peak electricity is expenditure inelastic.

Further, by employing an aggregate consumer behavior model which allows us to pool aggregate and micro data sets, we were able to estimate the influence of some interesting household characteristics' variables on time-of-use electricity demand. The estimation results show that households with larger capacity (largely the effect of the presence of a warm-water boiler) and large (four or more persons) households do substitute off-peak for peak electricity consumption.

Finally, the aggregate consumer behavior model represents a useful approach in estimating electricity demand by time-of-use by pooling aggregate data sets with micro data sets. However, data requirements for the estimation of this model are formidable. For instance, information on summary statistics reflecting the distribution of electricity expenditures across households, and summary statistics reflecting the distribution of household characteristics are available with limitations. This implies that a limited number of explanatory variables are usually included in the model specification.

CHAPTER 5

HOUSEHOLD CONSUMER BEHAVIOR MODEL

As presented in the previous chapter, the estimation of a demand system for electricity by pooling information contained in aggregate and micro data sets permits estimation of effects of price and household characteristics in ways that are preferable to those that are available using aggregate data sets only. However, the estimation of the aggregate consumer behavior model has the shorthcoming that the data needed at the aggregate level to estimate it are formidable, with the result of a limited number of household characteristics variables included in the model.

A possible solution to this problem, indicated in chapter 2, consists of the estimation of demand systems using cross-sectional or panel micro data sets that show price variation and contain accurate information on household-characteristics variables.

To verify the results in terms of price elasticities presented in the previous chapters and derived from the estimation of the representative and aggregate consumer models, in this chapter we estimate a demand system for electricity using a micro data set on Swiss households, which contains detailed information on household characteristics as well as enough price variation to estimate price elasticities. Therefore, the comparison of the estimation results reported in this chapter with those shown in the previous chapters will indicate if useful information on the behavior of individual households can be derived using the representative consumer model or the aggregate consumer model approach.

The chapter is organized as follows: In the first section, the household AIDS model for electricity is presented. In section 5.2, the micro data set employed in the analysis is discussed while in sections 5.3 and 5.4, the empirical results and estimates of the elasticities are presented. Some concluding remarks follow in section 5.5[61].

5.1 Model specification

The household AIDS demand model presented in section 4.1 is used in this section. The household AIDS model can be written as:

$$w_{ih} = \alpha_i + \sum_j \gamma_{ji} \ln p_j + \beta_i \ln(\frac{m_h}{P_h}) + \sum_d \eta_{id} c_{hd} \qquad (5.1)$$

where w_{ih} is the budget share of the ith good for the hth household, p_j are the prices, c_h is a vector with elements c_{hd} representing household characteristics, and (m_h/P_h) is the total real expenditure on all n goods in the consumers budget, with P_h a price index defined by

$$\ln P_h = \alpha_0 + \sum_j \alpha_j \ln p_j + \frac{1}{2}\sum_i \sum_j \gamma_{ij} \ln p_i \ln p_j + \sum_i \sum_d \eta_{id} \ln p_i c_{hd}$$

and $\gamma_{ij} = (\gamma^*_{ij} + \gamma^*_{ji})/2$. (5.2)

Similar to previous chapters, it is assumed that Stone's index provides an acceptable linear approximation to $\ln P_h$. Moreover, for

61 This chapter is based on a paper published in the *Energy Economics*. See Filippini (1995b).

consistency with demand theory, the following standard restrictions must hold:

Adding up: $$\sum_i \alpha_i = 1,\ \sum_i \gamma_{ij} = 0,\ \sum_i \beta_i = 0,\ \sum_i \eta_{id} = 0 \tag{5.3}$$

Homogeneity: $$\sum_j \gamma_{ij} = 0 \tag{5.4}$$

Symmetry: $$\gamma_{ij} = \gamma_{ji} \tag{5.5}$$

Following the previous model specifications, we assume that the utility of peak and off-peak electricity is weakly separable from the quantities consumed for all other goods. Moreover, the demographic effects are incorporated in the model by allowing the intercept in (5.1) to be a function of household characteristics represented as dummy variables, equal to 1 for a household having that characteristic and 0 otherwise. Therefore, the household budget shares of peak and off-peak electricity consumption are expressed as a function of electricity prices, real electricity expenditures and dummy variables representing household characteristics. As in the previous chapters we estimate both homothetic and non-homothetic versions of the conditional demand system for electricity.

The household characteristics variables hypothesized to impact consumption patterns of electricity are divided into four categories. The first category, called size characteristics, includes variables describing the size of the household. The second category, called member characteristics, is comprised of variables describing individuals within households, such as the occupation, age and sex of household members. The third category, called locational characteristics, is made up of variables describing the regional location of the household, such as the climate and the rate structure. The fourth category, called electrical appliances characteristics, includes variables describing electrical appliances utilized within the household, such as an electric boiler and a dishwasher. Moreover, for the majority of the electrical appliances considered in the analysis the time of use is discretionary.

For the reasons presented in the previous chapter, the stock of electrical appliances is again considered an exogenous household characteristic.

Thus, starting with equations (5.1) and assuming separability, the household AIDS consumer model for electricity consumption during the peak and off-peak periods can be written as

$$
\begin{aligned}
w_{hP} = {} & \mu_{hP} + \gamma_{PEP} \, lnP_{EP} + \gamma_{PEO} \, lnP_{EO} + \beta_P \, ln\frac{m_h}{P} + \eta_{PHS1} \, D_{HS1} \\
& + \eta_{PHS2} \, D_{HS2} + \eta_{PHS3} \, D_{DHS3} + \eta_{PR} \, D_R + \eta_{PHW} \, D_{HW} + \eta_{PC} \, D_C \\
& + \eta_{PDP} \, D_{DP} + \eta_{PHDD} \, D_{HDD} + \eta_{PDW} \, D_{DW} + \eta_{PW} \, D_W + \eta_{PF} \, D_F \\
& + \eta_{PWA} \, D_{WA} + \eta_{PD} \, D_D + \eta_{PK} \, D_K + \eta_{PHA} \, D_{HA}
\end{aligned} \tag{5.5}
$$

$$
\begin{aligned}
w_{hO} = {} & \mu_{hO} + \gamma_{OEP} \, lnP_{EP} + \gamma_{OEO} \, lnP_{EO} + \beta_O \, ln\frac{m_h}{P} + \eta_{OHS1} \, D_{HS1} \\
& + \eta_{OHS2} \, D_{HS2} + \eta_{OHS3} \, D_{DHS3} + \eta_{OR} \, D_R + \eta_{OHW} \, D_{HW} + \eta_{OC} \, D_C \\
& + \eta_{ODP} \, D_{DP} + \eta_{OHDD} \, D_{HDD} + \eta_{ODW} \, D_{DW} + \eta_{OW} \, D_W + \eta_{OF} \, D_F \\
& + \eta_{OWA} \, D_{WA} + \eta_{OD} \, D_D + \eta_{OK} \, D_K + \eta_{OHA} \, D_{HA}
\end{aligned} \tag{5.6}
$$

where w_{hP} = per household share of electricity expenditure during the peak period, obtained by dividing expenditures on peak electricity by total electricity expenditure, w_{hO} = per household share of electricity expenditure during the off-peak period, P_{EP} = electricity price during the peak period, P_{EO} = electricity price during the off-peak period, (m/P) = per household total expenditure on electricity divided by the Stone price index for electricity goods (P), D_{HS1} = dummy variable indicating whether the household is a one-person household, D_{HS2} = dummy variable indicating whether the household is a two or three persons household, D_{HS3} = dummy variable indicating whether the household is a four or more household, D_R = dummy variable indicating whether the household head is retired, D_{HW} = dummy variable indicating whether a member of the household is a housewife, D_C = dummy variable indicating whether the household has children, D_{HDD} = dummy variable indicating whether the household is living in a city characterized by more than 3300 heating degree days per year, D_{DP} = dummy variable indicating whether the electric utility considers the week-end an off-peak period, D_W = dummy variable indicating whether the household has an electric boiler, D_F = dummy variable indicating whether the household has a freezer, D_{WA} = dummy variable indicating whether the household has an electric washing machine, D_D = dummy variable indicating whether the household has a electric clothes-dryer, D_K = dummy variable indicating whether the household has an electric kitchenrange, D_{HA} = dummy variable indicating whether the household has small heating appliances and D_{DW} = dummy variable indicating whether the household has a dishwasher.

The expected impact of the household characteristics dummy variables on consumption patterns of electricity are the following:

- a small family is expected to consume less electricity during the peak period than a large family given a certain stock of appliances in the household;
- households living in an area with a high number of heating degree days are expected to have more need for space heating during the peak period than households living in cities located in the south of Switzerland where the climate is more temperate;
- households where at least one household member is retired, has a child or a housewife are expected to consume more electricity during the peak period than households with only workers;
- households living in cities where the weekend is considered an off-peak period are expected to have lower electricity consumption during the

peak period than households living in cities not affected by this tariff rule;

- households with an electric boiler are able to shift their water heating activities to an the off-peak period of electricity consumption. Therefore, these households are expected to have a higher share of off-peak electricity consumption than households without an electric boiler;

- households with an electric kitchen range are expected to have a higher share of peakelectricity consumption than households with a gas kitchen range;

- households with small electrical heating appliances or an electric washing machine, dishwasher or dryer are expected to have a higher share of peak electricity consumption than households without these electrical appliances because these appliances are generally used during the peak period.

5.2 Data and variables

The household micro data used in this chapter is provided in the first instance by the consumer demand surveys run in 1991 by the private marketing company IHA. The survey for the year 1991 contains data from 1,600 households living in Swiss cities. As in the case of households of Zurich we completed this data set with information from the electric utilities on the electricity consumption of each household in the survey. Of the 1,600 households sampled in 1991 by this marketing company, 387 did not provide us with permission to ask the electric utilities about their electricity consumption. Hence, the sample was restricted to 1,113 households. Of this 1,113 households sampled, 20% live in cities offering time differentiated tariffs to all customers while 80% live in cities offering time differentiated tariffs only to customers who use a lot of electricity during the off-peak period, or who decide to finance the additional metering costs for the application of a time-of-use tariff.

The sample of households employed to estimate equations (5.5) and (5.6) was restricted to the 217 households who live in the 19 Swiss cities of the sample offering time differentiated tariffs to all customers.[62]

Tables 5.1 and 5.2 give some details on the variables employed in the estimation of the household AIDS model for electricity.

Table 5.1 Description of the variables

household share equations variables	25th Percentile	Median	75th Percentile
Budget share of electricity consumption during peak-period (W_P)	0.65	0.79	0.85
Price during the peak period (P_{EP}) [in SwF/kWh]	0.16	0.17	0.19
Price during the off-peak period (P_{EO}) [in SwF/kWh]	0.07	0.08	0.09
Total electricity expenditure (m) [in SwF]	315	436	640

62 We restricted the sample to these 217 households because the empirical analysis of the electricity demand by time-of-use of households faced with self-selecting time-of-use tariffs needs a different theoretical and empirical approach than the one chosen in this study. Thus, we decided to leave this particular analysis for a future research. Moreover, using data on households living in cities where time-of-use tariffs are applied to all customers allows us to compare the empirical results with those obtained in previous chapters using analogous data sets.

Table 5.2 Description of sample household characteristics

Household characteristics	Variable	Condition for which the variable value is equal to one	Frequency (%)
Household size	DHS1	one person household	10
	DHS2	two or three person household	37
	DHS3	four and more person household	53
climate	DHDD	household living in a city with more than 3,300 heating degree days per year	80
occupation of the household	DR	household head retired	29
	DHW	housewife in household	46
household age composition	DC	household with children	27
peak and off-peak period	DDP	weekend is off-peak period	30
electrical appliances in household	DW	electric water heater (or electric boiler)	22
	DF	freezer	69
	DDW	dishwasher	40
	DWA	electric washing machine	44
	DD	electric clothes dryer	31
	DK	electric kitchen range	87
	DHA	small heating appliances	40

5.3 Estimation results

As in the previous models, because of the adding-up constraint, only the equation on peak electricity demand is estimated with the imposition of the conditions suggested by demand theory.

Table 5.3 presents the OLS estimated coefficients for the non-homothetic model. The statistical results are satisfactory, with most of the important coefficients having the expected signs and reasonable magnitude. Leaving discussion of the estimated price and expenditure elasticities for the next section, several interesting comments can be drawn from the remainder of the estimated coefficients shown in Table 5.3.

Fourteen household characteristics coefficients are estimated in the model. In Table 5.3 we see that 3 of these coefficients are significantly different from zero at the 95% confidence level and 1 of these coefficients at 90% confidence level. The results indicate that ceteris paribus, a small household uses less electricity during the peak period than a large household. Moreover, households with children appear to have a significantly higher budget share of peak electricity consumption than households without children, ceteris paribus.

With regard to the locational characteristics variables, we can observe that the budget share of peak electricity is generally lower for the households living in cities where the weekend is considered an off-peak period.

With the exception of the electric boiler, the estimated coefficients of the other electrical appliance variables do not have a significant influence on the budget share of peak electricity. This finding confirms the result obtained in the previous chapter that households with an electric boiler have a lower budget share of peak electricity than households without the electric boiler, because they can chose whether to turn on the electric boiler during the off-peak period and turn it off during the peak period of electricity consumption.

The coefficient on real electricity expenditure is significantly different from zero. This result confirms that homotheticity is not a valid assumption in estimating electricity demand by time-of-use.

Table 5.3 Estimates of the non-homothetical time-of-use share equation (standard errors in parentheses)

Independent Variables	OLS (&) Dependent variables w_P	OLS Dependent variables w_P
constant	1.557 ***	1.557 ***
	(0.208)	(0.141)
$\ln P_{EP}$	-0.259 **	-0.259 **
	(0.123)	(0.125)
$\ln P_{EO}$	0.259 **	0.259 **
	(0.123)	(0.125)
ln (m/P)	-0.069 ***	-0.069 ***
	(0.025)	(0.014)
DHS1	-0.050 *	-0.050 **
	(0.030)	(0.024)
DHS3	-0.021	-0.021
	(0.017)	(0.017)
DHDD	-0.011	-0.011
	(0.018)	(0.020)
DR	-0.004	-0.004
	(0.016)	(0.017)
DHW	-0.001	-0.001
	(0.013)	(0.014)
DC	0.061 ***	0.061 ***
	(0.019)	(0.018)
DDP	-0.141***	-0.141***
	(0.016)	(0.016)
DW	-0.174 ***	-0.174 ***
	(0.019)	(0.017)
DF	0.021	0.021
	(0.016)	(0.014)
DDW	-0.018	-0.018
	(0.014)	(0.014)
DWA	-0.006	-0.006
	(0.012)	(0.015)
DD	0.014	0.014
	(0.015)	(0.015)
DK	0.020	0.020
	(0.022)	(0.020)
DHA	0.009	0.009
	(0.013)	(0.014)
Adjusted R^2	0.59	0.59

*, **, ***: significantly different from zero at the 90%, 95%, 99% confidence level; (&) The standard errors presented for this OLS model are calculated from (White's) heteroschedastic-consistent variance-covariance matrix.

Table 5.4 Estimates of the homothetical time-of-use share equation (standard errors in parentheses)

Independent Variables	OLS (&) Dependent variables w_P	OLS Dependent variables w_P
constant	1.124 ***	1.124 ***
	(0.114)	(0.114)
$\ln P_{EP}$	-0.332 ***	-0.332 **
	(0.127)	(0.131)
$\ln P_{EO}$	0.332 ***	0.332 **
	(0.127)	(0.131)
DHS1	-0.023	-0.023
	(0.028)	(0.025)
DHS3	-0.034*	-0.034*
	(0.018)	(0.018)
DHDD	-0.027	-0.027
	(0.018)	(0.021)
DR	-0.005	-0.005
	(0.017)	(0.018)
DHW	-0.004	-0.004
	(0.013)	(0.015)
DC	0.059 ***	0.059 ***
	(0.020)	(0.019)
DDP	-0.144 ***	-0.144 ***
	(0.017)	(0.016)
DW	-0.195 ***	-0.195 ***
	(0.018)	(0.017)
DF	0.013	0.013
	(0.016)	(0.015)
DDW	-0.030 **	-0.030 **
	(0.015)	(0.015)
DWA	-0.027 *	-0.027 *
	(0.016)	(0.015)
DD	0.008	0.008
	(0.017)	(0.016)
DK	-0.004	-0.004
	(0.021)	(0.021)
DHA	0.012	0.012
	(0.015)	(0.014)
Adjusted R^2	0.55	0.55

*, **, ***: significantly different from zero at the 90%, 95%, 99% confidence level; (&) The standard errors presented for this OLS model are calculated from (White's) heteroschedastic-consistent variance-covariance matrix.

Table 5.4 shows estimated coefficients from the homothetical time-of-use share equation regression analysis. The results confirm that price, electric boiler, children and the duration of the off-peak period have a significant impact on the budget share of peak electricity. However, the values of the estimated coefficients are generally different from those found in the non-homothetical specification. These differences could be caused by the omission from the model of an important explanatory variable such as the total electricity expenditure. Because electricity expenditure is correlated with many included variables, the estimated values of the regression coefficients reported in Table 5.4 are biased. Therefore, the estimates of the homothetical time-of-use share equation presented in this and in the previous two chapters have to be interpreted carefully.

5.4 Partial elasticity estimates

Table 5.5 presents the partial own price, cross price, and substitution elasticities for electricity demand by time-of-use evaluated at the median budget share for both homothetic and non-homothetic household consumer model specification.

Table 5.5 Estimated partial elasticities

	Non-homothetic model	Homothetic model
Price elasticity,peak	-1.25	-1.41
Price elasticity, off-peak	-2.30	-2.57
Cross price elasticity, peak/off-peak	0.34	0.41
Cross price elasticity, off-peak/peak	0.97	1.57
Elasticity of substitution peak/off-peak	2.56	2.98

The results in terms of elasticities presented in Table 5.5 confirm the results obtained by the estimates of the representative consumer behavior model and the aggregate consumer behavior model, i.e., they have the same sign and similar values. The estimated own price elasticities vary from -1.25 to -1.41 during the peak period and from -2.30 to -2.57 during the off-peak period. These elasticities show high responsiveness of electricity consumption to changes in peak and off-peak prices. Moreover, the positive values of the elasticity of substitution confirm that peak and off-peak electricity are substitutes. The AUES matrix evaluated at the mean budget shares is negative semi-definite and, the results are, therefore, in accord with consumer theory.

Finally, the estimated partial expenditure elasticities are 0.91 for peak electricity and 1.33 for off-peak electricity, confirming the results obtained in chapter 3 that off-peak electricity is expenditure elastic and peak electricity is expenditure inelastic.

5.5 Concluding comments

In this chapter, we estimated the price and expenditure elasticities of peak and off-peak electricity consumption using a micro data set on 217 households living in 19 Swiss cities. The household version of the Almost Ideal Demand System model was used as a framework. To incorporate household characteristics, the AIDS model was expanded by specifying the intercept as a linear function of household-characteristics variables. Results indicate that demand for peak and off-peak electricity is elastic. Cross-price effects are positive as well as the elasticity of substitution. Therefore, peak and off-peak electricity are substitutes in the Hicks-Allen sense. These results support the values of price and substitution elasticities found in previous chapters by estimating the representative and the aggregate consumer behavior models. Moreover, the results confirm that the nonhomothetical specification has to be preferred to the homothetical one.

The most significant household characteristic affecting the share of peak and off-peak electricity consumption come from household size and from the presence of children in households. Moreover, the estimation results show that the presence of an electrical warm water boiler, whose time of use is discretionary, strongly influences the electricity consumption pattern.

CHAPTER 6

SUMMARY AND CONCLUSIONS TO PART I

This chapter presents some insights concerning the implementation of demand theory in analyzing electricity demand by time-of-use, important empirical findings, and several policy implications.

Insights concerning the implementation of demand theory

The main objective of this part of the book was to provide empirical evidence on the residential demand for electricity by time-of-use in Switzerland. For this purpose, we assumed that consumer demand for electricity, other energy and non-energy commodities is the result of a two-stage budgeting process. First the consumer determines the amount to spend on electricity, other energy and non-energy commodity groups based on price indexes for each group.Then, the consumer allocates this electricity budget among peak and off-peak electricity commodities. The first-stage demand relation was not estimated in this study.

Because different data sets were available, three types of consumer models have been employed. The first is the representative consumer behavior model suggested by Muellbauer (1975, 1976). The second is the aggregate consumer behavior model proposed by Jorgenson, Lau and Stocker (1982) and based on the exact aggregation theory of Lau (1982). The third model is a classical household consumer behavior model.

Moreover, all models were estimated using different versions of the Almost Ideal Demand System (AIDS) functional form proposed by Deaton and Muellbauer (1980a).

To analyze electricity demand during peak and off-peak periods using the representative consumer model, panel data covering four years (1987,1988,1989,1990) and 21 cities in Switzerland were used. This aggregate panel data set shows a greater cross-section or panel variation in electricity price variables, which permitting a more accurate estimation of the price coefficients. However, this data set has the disadvantage of containing inaccurate and limited information on household characteristics which are usually well documented in micro data sets. Therefore, the results obtained by the estimation of the representative consumer model presented in chapter 3 have two shortcomings. First, they do not contain information on the effects on electricity consumption by time-of-use of household characteristics variables such as the electrical appliance stock. Second, because of imprecise information on the household characteristics variables included in the model, the estimated coefficients of these variables may also be inexact.

A first approach to overcome these difficulties was to estimate the aggregate consumer behavior model proposed by Jorgenson, Lau and Stocker (1982) by pooling micro and aggregate data sets. The micro data set utilized contains data for the years 1989-1990 on 280 households living in Zurich. Because this data set, as most micro data sets, refers to a sample of households faced with the same prices, we could not estimate elasticities of price and substitution. However, using the aggregate consumer behavior model, we were able to combine the aggregate data set with this micro data set allowing us the estimation of the coefficients on prices and some household-characteristics variables of interest. Although the estimation of the aggregate consumer behavior model by pooling micro and aggregate data sets represents an improvement over the representative consumer behavior model, it has to be noted that the rather formidable data requirements needed to estimate it with aggregate panel data, limited the number of household-characteristics variables introduced in the model. However, from a theoretical and empirical point of view we consider this model appealing, because it exploits information from data sets on more levels and provides a theoretically sound link between household consumer and aggregate consumer behavior.

Finally, to analyze in more detail the influence of various household characteristics variables on residential electricity consumption patterns, we estimated a household consumer behavior model by exploiting information contained in a micro data set on 217 households living in 19 Swiss cities for the year 1991. This micro data set has the advantage of showing price variation as well as including precise information on numerous variables reflecting household characteristics.

The empirical analysis proposed in this part of the book adds to the existing literature in four respects: First, it presents and compares the estimation results of a demand system for electricity using different data sets and three different specifications of consumer behavior models. Second, it combines information pertaining to different levels of aggregation, individual households and cities. Our results indicate that using the exact aggregation theory of Lau (1982), it is feasible to estimate price, expenditure and household-characteristics effects in a demand system by pooling aggregate panel data and disaggregate panel data. Moreover, the application of the exact aggregation theory in modelling aggregate demand is attractive since it provides a link between the consumer demand theory and its empirical application. Third, our empirical analysis shows that the estimation of the household consumer behavior model is preferable to the estimation of the representative consumer and aggregate consumer behavior models because it allows a more precise estimation of the effects of household characteristics on electricity demand by time-of-use. Fourth, whereas the majority of studies on time-of-day pricing of electricity have their empirical basis in experiments where consumers were faced with a variety of schedules and prices with, however, the guarantee that they would not have to pay more than the usual amount for their electricity consumption, the empirical part of this paper is based on observational data.

Empirical findings

The most important objective of this part of the book was to determine price and substitution elasticities of demand for electricity by time-of-use. Moreover, using three different data sets we wanted to compare the estimation results of the representative, aggregate and household consumer behavior models.

With regard to choice of econometric technique for the estimation of both aggregate and representative consumer behavior models with panel data, the following regression models were selected: OLS, least squares dummy variable (LSDV) , and error components (EC). Because the between variation for the relevant prices explanatory variables in the model (i.e., the variation of the electricity prices between the cities) greatly exceeds the within variation (i.e., the variation over time), the OLS becomes a preferred estimator. Thus, following the arguments of Baltagi and Griffin (1984), price and substitution elasticities are based on the OLS estimates results.

The estimation results show that the price and substitution elasticities of all models are similar in sign and magnitude. Moreover, the estimation results of all models reject the homotheticity assumption. However, while the results of the aggregate consumer model indicate that peak electricity is expenditure elastic and off-peak electricity is expenditure inelastic, those of the representative and household consumer behavior models suggest the opposite, e.g., peak electricity is expenditure inelastic and off-peak electricity is expenditure elastic. This dissimilarity in estimating electricity expenditure elasticities can also be found in previous studies on peak and off-peak electricity demand. This difference may be attributed to the way in which peak and off-peak period are implemented in the different data sets. For example, to estimate the household consumer behavior model we employed a data set of households living in Swiss cities where the weekend is either considered as an off-peak period or a mixed peak and off-peak period. Whereas in estimating the aggregate consumer behavior model we combine a data set of households living in Zürich, where the weekend is considered a mixed peak and off-peak period, with an aggregate data set pertaining to Swiss cities where the weekend is either considered as an off-peak period or a mixed peak and off-peak period. Future research should, therefore, take into account the impact of days of on-peak and off-peak rates on electricity expenditure elasticities.

Table 6.1 presents the estimated elasticities derived from the estimation results of the three consumer behavior models.

Table 6.1 Summary of the estimated partial elasticities

	Representative consumer behavior model Non-homothetic model	Aggregate consumer behavior model Non-homothetic model	Household consumer behavior model Non-homothetic model
Price elasticity, peak	-1.39	-1.58	-1.25
Price elasticity, off-peak	-2.58	-2.54	-2.30
Cross price elasticity, peak/off-peak	0.50	0.54	0.34
Cross price elasticity, off-peak/peak	1.24	1.67	0.97
Elasticity of substitution peak/off-peak	2.97	3.13	2.56
Expenditure elasticity peak	0.89	1.05	0.91
Expenditure elasticity off-peak	1.33	0.86	1.33

These results can be summarized as follows:

- the estimated own-price elasticities of peak and off-peak electricity consumption have the expected sign and indicate a high responsiveness of electricity consumption to changes in peak- and off-peak prices;
- the positive values of the cross price elasticities and of the elasticity of substitution show that peak and off-peak electricity are substitutes. Therefore, our results support the hypothesis that electricity time-of-use rates can have the effect of shifting residential electricity consumption from peak to off-peak hours;

- the estimation results of all models reject the homotheticity assumption of the electricity demand by time-of-use.

In addition, from the estimation of the aggregate and household consumer models the following general findings concerning the effects on household characteristics on the electricity demand consumption pattern were discovered:

- the presence of retired people in a household or a housewife in a household did not influence the electricity consumption pattern, but the number of household members and the presence of children in a household did;
- freezer, dishwasher, clothes washer, electric dryer and small electric heating appliances ownership did not have a significant impact on usage pattern, but electric water-heating ownership did.

The estimated price elasticities are higher than those found in other studies. However, a comparison of the results is difficult, and typically inconclusive since models, data, and time periods used are not similar.

Finally, the implications for the applied economist resulting from the empirical analysis presented in this part of the book are:

- in terms of price and substitution elasticities the estimation results of the representative and aggregate consumer behavior models suggest that using these models it is possible to estimate parameters of micro demand behavior from aggregate data;
- pooling aggregate data with micro data to estimate a model of consumer behavior is a useful solution when micro data sets do not show enough price variation to estimate price and substitution elasticities;
- for a precise analysis of the impact of household characteristics variables on the electricity consumption pattern, the household consumer model is useful than the representative or aggregate consumer behavior models.

Policy implications

The empirical results on the residential electricity demand by time-of-use presented in this part of the book suggest answers to the energy policy issues described in the first chapter of this study. Moreover, these results have implications that may be exploited by Swiss electric utilities managers and industry representatives.

From the standpoint of conserving end-use electricity, it is of great importance to know peak and off-peak demand elasticities with respect to individual electricity prices. The fact that own price elasticities are all negative and that substitution elasticities are all positive has an important implication for conservation. It suggests that pricing policy can be an effective instrument for achieving electricity conservation and that time-of-use pricing in particular can contribute to more efficient utilization of existing production and distribution capacity.

Thus, to reduce peak-period electricity consumption in Switzerland we suggest that utilities which already apply time-of-use rates, increase the differential between peak and off-peak price, whereas utilities that do not yet apply time-of-use rates in the residential sector, should introduce it.

Reductions in peak-period demand will permit utilities to run existing electric plants efficiently and to reduce the steady growth of peak electricity demand, which is the chief determinant of the need for new capacity. This would allow the buildup of additional capacity to be postponed. Moreover, as will be shown in the second part of this book, we found that an improvement in the load factor, which can be influenced by time-of-use rates, can significantly reduce electric power distribution costs. In addition, widespread adoption of time-of-use rates in the Swiss residential sector could help to achieve an optimal domestic provision of electricity during the winter in a market with trade.

PART II

THE COST STRUCTURE OF SWISS ELECTRIC UTILITIES

INTRODUCTION TO PART II

The purpose of part II of this book is to study the cost structure of the Swiss electricity distribution utilities. The principal aims of the following chapters are to determine what factors affect cost, and the nature of economies of scale in this industry. Moreover, since electricity is a non-storable good with a cyclic demand that varies in the course of the day, the impact of different daily system load curves on electricity distribution costs is presented.

To develop a better understanding of the cost structure of the Swiss electricity distribution industry, both long- and short-run equilibrium approaches are used. We specify total and variable cost functions as a translog form and estimate them using panel data on 39 electricity distribution utilities during the period 1987 to 1991.

The empirical evidence presented in this part of the book is relevant to several regulatory issues. First, it presents information about the validity of the natural monopoly argument in the distribution of electricity and, therefore, of the common justification for the distribution of electricity by local monopolies. Second, it contributes to an evaluation of the desirability of competition in the distribution of electric power. Third, it fills an important gap in the literature on electricity cost studies by estimating total and variable cost functions for electricity distribution utilities. Finally, it provides empirical evidence on the impact on costs of the shape of the system load curve and, therefore, of the potential impact of changes in time-of-use rates on electricity distribution costs.

This part of the book is organized into four chapters. Chapter 7 discusses the problems and issues related to implementing producer theory. Chapter 8 presents an empirical analysis of the long-run cost model approach, while in chapter 9, the results of the estimation of the short-run cost model are discussed. Summary and conclusions are presented in chapter 10.

CHAPTER 7

IMPLEMENTING PRODUCER THEORY

This chapter introduces several elements useful for applied economists by implementing the theory of producer behavior. The first section reviews traditional production theory as well as concepts developed under the duality theory as applied to analysis of producer behavior. In the second section we present the translog functional form, which is used in chapters 8 and 9 to estimate the long- and short-run cost functions for the Swiss electric distribution utilities. The third section presents possible ways to account for heterogeneity of output and technological environment of the firm in a cost model. Finally, a survey of the most important econometric analysis of electricity cost functions is discussed in the last section of this chapter.

7.1 Review of traditional production theory

The microeconomic theory of production is well documented in the literature, e.g., Shephard, R.W. (1953), Chambers (1988), Jehle (1991) and Varian (1992), and it is not the purpose of this section to repeat this material in detail. Instead, this section focuses on some elements of the microeconomic theory of production that are relevant for the development of a cost model for the Swiss electricity distribution industry.

Long- and short-run cost functions

We model the production of firms in an industry which use $g+c$ inputs $\mathbf{x} = (x_1, x_2, ..., x_{g+c})$ to produce m outputs $\mathbf{y} = (y_1, y_2, ..., y_m)$. A reasonable way to represent the firm's technology of turning inputs into output in the long run is to specify a transformation function, $T(x_1, ..., x_{g+c}, y_1, ..., y_m) = 0$, in the multiple output case or as a production function, $y = f(x_1, x_2, ..., x_{g+c})$ in the single output case.

If the firm faces competitive input markets and chooses input bundles to minimize costs in the long run, then the cost minimizing process can be represented as

$$\begin{aligned} &\min_{\mathbf{x}} C = \sum_{j=1}^{g+c} w_j x_j \\ &\text{s.t. } f(\mathbf{x}) \geq \mathbf{y} \end{aligned} \tag{7.1}$$

where C represents long-run total cost, w_j is the price of input x_j, and f is the production function relating the vector of inputs $\mathbf{x}$ to the output vector $\mathbf{y}$. The solution to (7.1) is of the form $C(\mathbf{y}, \mathbf{w})$, where $\mathbf{y} = (y_1, y_2, ..., y_m)$ and $\mathbf{w} = (w_1, w_2, ..., w_{g+c})$, as before.

Given that the transformation function $T(x_1, ..., x_{g+c}, y_1, ..., y_m) = 0$ has a strictly convex input structure, McFadden (1978) has shown that the cost function, $C(\mathbf{y}, \mathbf{w})$ has the following properties (regularity conditions):

1. $C(\mathbf{y}, \mathbf{w}) > 0$ for $\mathbf{w} > 0$ and $\mathbf{y} > 0$ (non-negativity)

2. if $\mathbf{w}' > \mathbf{w}$, then $C(\mathbf{y}, \mathbf{w}') \geq C(\mathbf{y}, \mathbf{w})$ (non-decreasing in $\mathbf{w}$)

3. concave and continuous in $\mathbf{w}$

4. $C(\mathbf{y}, \mathbf{w})$ is homogeneous of degree one in input prices: $C(\mathbf{y}, t\mathbf{w}) = t\,C(\mathbf{y}, \mathbf{w})$ for $t > 0$

5. if $\mathbf{y} > \mathbf{y}'$, then $C(\mathbf{y}, \mathbf{w}) \geq C(\mathbf{y}', \mathbf{w})$ (non-decreasing in $\mathbf{y}$)

6. $C(0, \mathbf{w}) = 0$ (no fixed costs).

Additionally, according to Shephard's lemma, the cost minimizing input demand functions are derivable from the cost functions. If the cost function is differentiable in input prices at the point (**y***, **w***) then the following property holds:

7. Shephard's lemma : $\mathbf{x}(\mathbf{y},\mathbf{w}) = \nabla_{\mathbf{w}} C(\mathbf{y}^*,\mathbf{w}^*)$

   where **x** denotes the cost-minimizing vector of inputs required to produce the **y*** vector of outputs given input prices **w***.

If we desire to represent the firm's technology of turning inputs into output in the short run, where the firm is committed to employing certain amounts of at least one factor, the transformation function becomes $T(x_1, \ldots, x_g, y_1, \ldots, y; k_1, \ldots, k_c) = 0$, where $\mathbf{y} = (y_1, y_2, \ldots, y_m)$ as before, $\mathbf{w_g} = (w_1, w_2, \ldots, w_g)$ contains prices of perfectly variable inputs, and $\mathbf{k} = (k_1, k_2, \ldots, k_c)$ is the quasi-fixed input vector.

The long-run total cost methodology share the stringent behavioral assumption that all inputs are employed at their long-run cost-minimizing level. In the following, we present a model where, in the short run, some inputs available to the firms are assumed to be fixed, implying that firms attempt to minimize cost conditional on given quasi-fixed inputs.

If the firm faces competitive input markets and chooses input bundles to minimize costs in the short run, then the cost minimizing process can be represented as

$$\begin{aligned} &\min_{\mathbf{x}} \ VC = \sum_{j=1}^{g} w_j x_j \\ &\text{s.t.} \ \ f(\mathbf{x},\mathbf{k}) \geq \mathbf{y} \end{aligned} \tag{7.2}$$

where VC represents short-run variable costs, w_j is the price of input x_j, and f is the production function relating the vectors of inputs **x** and quasi-fixed inputs **k** to the output vector **y**.

The solution to (7.2) is of the form VC(**y**, $\mathbf{w_g}$, **k**), where $\mathbf{y} = (y_1, y_2, \ldots, y_m)$, $\mathbf{w_g} = (w_1, w_2, \ldots, w_g)$ and $\mathbf{k} = (k_1, k_2, \ldots, k_c)$, as before.

Following Lau (1976) and Chambers (1988), if the firm minimizes the variable costs of producing a given output, subject to quasi-fixed inputs, the variable cost function, $VC(\mathbf{y}, \mathbf{w}_g, \mathbf{k})$ has the following properties (regularity conditions):

1. $VC(\mathbf{y}, \mathbf{w}_g, \mathbf{k}) > 0$ for $\mathbf{w} > 0$ and $\mathbf{y} > 0$ (non-negativity)

2. if $\mathbf{w}_g' > \mathbf{w}_g$, then $VC(\mathbf{y}, \mathbf{w}_g', \mathbf{k}) \geq VC(\mathbf{y}, \mathbf{w}_g, \mathbf{k})$ (non-decreasing in $\mathbf{w}_g$)

3. concave and continuous in $\mathbf{w}_g$

4. $VC(\mathbf{y}, \mathbf{w}_g, \mathbf{k})$ is homogeneous of degree one in input prices: $VC(\mathbf{y}, t\mathbf{w}_g, \mathbf{k}) = t\, VC(\mathbf{y}, \mathbf{w}_g, \mathbf{k})$ for $t > 0$

5. if $\mathbf{y} > \mathbf{y}'$, then $VC(\mathbf{y}, \mathbf{w}_g, \mathbf{k}) \geq VC(\mathbf{y}', \mathbf{w}_g, \mathbf{k})$ (non-decreasing in $\mathbf{y}$)

6. if $\mathbf{k}' \geq \mathbf{k}$ then $VC(\mathbf{y}, \mathbf{w}_g, \mathbf{k}') \leq VC(\mathbf{y}, \mathbf{w}_g, \mathbf{k})$ (non-increasing in $\mathbf{k}$)

Property 6 indicates that short-run variable costs should be non-increasing in constrained inputs. This implies that marginal increases in quasi-fixed inputs should allow the reduction of short-run variable costs.[63] As discussed in more detail in the following sections, several empirical studies which estimate translog variable cost functions failed to confirm property 6.

As in the case of the long-run cost function, by Shephard's lemma, the short-run cost minimizing input demand functions are derivable from the variable cost functions.

We have defined the short-run variable cost function as the minimum cost to produce a given level of output, only adjusting the variable inputs. Moreover, we define the long-run total cost function as the minimum cost of producing a given level of output, adjusting all of the inputs. We can now define the short-run total cost function as the sum of variable costs and fixed costs. The idea implicit in these short-run total costs is that they are

---

63 For a graphical presentation of property 6, see Chambers (1988).

the best that can be achieved if one is committed to **k**. The short-run total cost function (SRC) associated with producing **y** is

$$SRC(\mathbf{y}, \mathbf{w}, \mathbf{k}) = VC(\mathbf{y}, \mathbf{w_g}, \mathbf{k}) + \mathbf{w_k k} \tag{7.3}$$

where the input prices vector **w** is partitioned into two components, $\mathbf{w_g}$ containing prices of perfectly variable inputs while $\mathbf{w_k}$ containing prices of quasi-fixed inputs.

Clearly, by definition of the long-run total cost function (7.1) and of the short-run total cost function (7.3), we must have

$$C(\mathbf{y}, \mathbf{w}) \leq VC(\mathbf{y}, \mathbf{w_g}, \mathbf{k}) + \mathbf{w_k k} \tag{7.4}$$

with equality in (7.4) when for a given level of outputs the long-run cost-minimizing inputs combination **k*** is reached. Equality means that the minimum costs when all factors are variable is equal to the minimum costs when **k** is fixed at level **k*** that minimizes long-run costs. Therefore, the relationship between SRC(**y**, **w**, **k**) and C(**y**, **w**) is that for a given level of outputs, long-run total costs can never be greater than short-run total costs.

The long-run equilibrium conditions for cost minimization requires that (7.4) be minimized in terms of **k**, the choice variables, since **y** and $\mathbf{w_g}$ are assumed to be exogenous and VC has already been minimized in terms of **y**, $\mathbf{w_g}$ and **k**. Thus if $\mathbf{k}^*$ denotes the vector of the optimal value of the quasi-fixed inputs which minimize SRC, then at $\mathbf{k} = \mathbf{k}^*$,

$$\left.\frac{\partial SRC}{\partial k_j}\right|_{k=k^*} = \left.\frac{\partial VC}{\partial k_j}\right|_{k=k^*} + w_j^k = 0, \quad \text{or}$$
$$-\left.\frac{\partial VC}{\partial k_j}\right|_{k=k^*} = w_j^k. \tag{7.5}$$

Equation (7.5) represents the first-order conditions associated with (7.4) and implies that in the long-run equilibrium the reduction in variable costs resulting from the last unit of a quasi-fixed input equals the price of the quasi-fixed input.

There is a dual relationship between the cost functions and the transformation functions presented above. This implies that if we define a production function and derive its cost function, we can take that cost function to generate a production function. Further, any short-run variable cost function or long-run cost function satisfying properties 1-7 implies some technology for which it is the cost function. Therefore, according to the Shepherd Duality Theorem (1953), all structural characteristics of production possibilities are contained in the functional specification of the cost function satisfying properties 1-7.

In econometric applications, working with a cost function has many advantages over its dual transformation or production function. For instance, cost functions in comparison to transformation and production functions focus more on economic relationships than on technological relationships, thus facilitating the discussion of economic relationships. Moreover, the production function is typically not estimated because a high correlation between the input variables can result in a multicollinearity problem, which might preclude the estimation of the model or lead to an inprecise interpretation of the estimated coefficients. This problem is less pronounced when a cost function approach is employed. However, two assumptions are required for the cost function to be a valid dual representation of production technology. First, firms should face exogenous input prices and outputs. Second, firms are assumed to adjust input levels to minimize costs.

7.2 Functional form and the translog model

In the literature, a high number of alternative functional forms have been employed to estimate cost functions.[64] Generally, we can distinguish between traditional and flexible functional forms. The former are first-order differentiable approximations to an arbitrary continuously twice differentiable function, whereas the latter provide a second-order

---

64 For a good overview of functional form in applied production analysis see Chambers (1988) and Griffin et al. (1987).

differentiable approximation.[65] The most important difference between traditional and flexible functional forms is that the former impose restrictions on the values of the first and second partial derivatives whereas the latter do not. Traditional functional forms such as the Cobb-Douglas and the CES impose different a priori restrictions on technology. For example, the Cobb-Douglas form imposes the restriction of a unitary elasticity of substitution and a value of economies of scale that does not vary with output. Whereas flexible functional forms such as the translog allows for values of economies of scale, input price and substitution elasticities vary with output. Further, flexible functional forms allow the test of several technological characteristics, such as a homothetical technology.

Table 7.1 presents a review of some common functional forms employed in cost studies of the electric industry. The first empirical studies focused on the simple relationships between output and costs.[66] These studies employed uncomplicated cost models assumed to be linear or polynomial in output. Later, cost function models were based on the application of duality theory to neo-classical production theory utilizing more sophisticated functional forms. Nerlove's pioneering study on the cost structure of US electric utilities employed the traditional Cobb-Douglas specification.[67] The publication in 1971 of the paper of Christensen, Jorgenson and Lau on translog functional form strongly affected the applied production analysis literature. Apart from a small number of studies, all analysis on cost structures of electric utilities published in the last two decades made use of the translog functional form.

---

65 Within a group of flexible functional forms we can differentiate those functional forms derived from second-order Taylor series approximations, such as the translog, the generalized Leontief and the quadratic or the functional forms derived from Fourier or Laurent series' approximations. The former are characterized by local flexibility, which implies a perfect approximation for an arbitrary function and its first two derivatives at a particular point. The latter are distinguished by their global flexibility. In applied work, however, the global approximation in Fourier or Laurent form is not commonly used because estimation involves considerably more parameters than are required for a functional form derived from second-order Taylor series' approximations.

66 See for example Nordin (1947), Johnston (1952) and Huettner and Landon (1974).

67 See Nerlove(1963).

Table 7.1 Some common functional forms employed in cost studies of the electric industry

Function	*Functional form*	*First derivative*	*Linearly homogeneous*	*Homothetic*	*Constant elasticity of substitution*	*Constant value of economies of scale*	*Number of distinct parameters*
Linear	$C = \alpha + \sum_i \beta_i x_i$	unrestricted in sign but constant	with restrictions imposed	yes	yes	yes	n+1
Cobb-Douglas	$C = \alpha \prod_i x_i^{\beta_i}$	unrestricted in sign but non-switching in sign	with restrictions imposed	yes	yes	yes	n+1
Translog	$\ln C = \alpha + \sum_i \beta_i \ln x_i + \sum_i \sum_j \gamma_{ij} (\ln x_i)(\ln x_j)$	unrestricted in sign	no, unless restrictions imposed	no, unless restrictions imposed	no, unless restrictions imposed	no	1/2 (n+1)(n+2)
Quadratic	$C = \alpha + \sum_i \beta_i x_i + \sum_i \sum_j \gamma_{ij} x_i x_j$	unrestricted in sign	no, unless restrictions imposed	no, unless restrictions imposed	no, unless restrictions imposed	no	1/2 (n+1)(n+2)

C = short-run variable cost or long-run total cost

x_i = input prices and outputs, and/or fixed inputs

This functional form can be interpreted as a second-order Taylor series approximation of an arbitrary well-behaved cost function.[68] Starting, for example, from the following long-run cost function

$$C = C(y_1, y_2, ..., y_S; w_1, w_2, ..., w_P) \quad (7.6)$$

where C is total cost, w_1, w_2, ..., w_P are the input prices and y_1, y_2, ..., y_S are the outputs, the following translog cost function can be defined:

$$\ln C = \alpha_0 + \sum_{i=1}^{p} \alpha_i \ln w_i + \sum_{m=1}^{s} \beta_m \ln y_m + \frac{1}{2}\sum_{i=1}^{p}\sum_{j=1}^{p} \gamma_{ij} \ln w_i \ln w_j$$
$$+ \frac{1}{2}\sum_{m=1}^{s}\sum_{n=1}^{s} \mu_{mn} \ln y_m \ln y_n + \sum_{i=1}^{p}\sum_{m=1}^{s} \lambda_{im} \ln w_i \ln y_m \quad (7.7)$$

with $\gamma_{ij} = \gamma_{ji}$ and $\mu_{mn} = \mu_{nm}$.

As stated in section 1, a cost function can be used to represent a technology if the following regularity conditions hold: non-decreasing in input prices; concave in input prices; linear homogeneity in input prices; and non-decreasing in outputs. Usually, the condition of linear homogeneity in input prices is imposed a priori on estimation, while other restrictions have to be verified after estimation because no parametric restrictions ensure their validity. There are two ways to impose the restrictions of linear homogeneity. The first approach implies the following parametric restriction:

$$\sum_{i=1}^{p} \alpha_i = 1; \ \sum_{j=1}^{p} \gamma_{ij} = 0 \ \text{ for all } i, \text{ and } \sum_{i=1}^{p} \lambda_{im} = 0 \ \text{ for all } m \quad (7.8)$$

while the second uses one of the input prices to normalize cost and input price.

68 See Christensen, Jorgenson and Lau (1971) for a lucid presentation of the translog functional form and Berndt (1991) for a detailed presentation of the econometric implementation of this functional form.

As the translog form provides a second-order approximation of the true cost function at a given point, the independent variables are expressed as deviations from this point. Often, the sample mean or median of each of the cost function variables is chosen. Because of this, the intercept represents an estimate of cost at the point of approximation.

The translog cost function is a second-order approximation of the true cost function which agrees with its first and second derivatives at the point of approximation. For points far from the approximation point, for example points far from the sample data, the theoretical restrictions representing technology might not hold.[69]

The efficiency of parameter estimates can be improved by estimating the cost function along with the factor share equations implied by Shephard's Lemma. Because, according to theory, certain parameters in the cost function are identical to certain parameters in the share equations, additional degrees of freedom are gained without the need to estimate any additional parameters. The input share equations take the following form:

$$S_i = \alpha_i + \sum_{j=1}^{p} \gamma_{ij} \ln w_j + \sum_{m}^{s} \lambda_{im} \ln y_m \tag{7.9}$$

The translog functional form continues to be attractive in studies on cost structure of electric utilities for the following reasons:

- it allows to satisfy (at least locally) the regularity conditions implied by neo-classical production theory;
- it accommodates tests for a number of production characteristics such as homotheticity, constant elasticity of substitution, and constant value of economies of scale;
- in comparison with other flexible functional forms derived from Laurent and Fourier approximations, it involves the estimation of considerably fewer parameters;

69 See Fuss, McFadden and Mundlak (1978).

- it allows for the values of economies of scale, input price, and substitution elasticities to vary with output;
- in comparison with other flexible functional forms derived from second-order Taylor series approximations, its first order coefficients are directly interpretable as cost elasticities evaluated at the approximation point;

However, it has to be observed that in comparison to traditional functional forms, the use of flexible functional form such as the translog, which are characterized by numerable interaction terms, has the disadvantage of increasing the possibility of multicollinearity between explanatory variables. If the purpose of the model is to predict, then multicollinearity is irrelevant, but if the purpose is to estimate technological characteristics, than a high degree of correlation between the explanatory variables may give rise to problems in the interpretation of the coefficients.

In the literature, the problem of multicollinearity in flexible functional forms has been theoretically adressed by Fuss, McFadden and Mundlak (1978), while the empirical results of Guyomard and Vermersch (1989) and Filippini (1991) show that when estimating translog variable cost functions, correlation between the explanatory variables, „output" and „capital stock" commonly causes a multicollinearity problem. Moreover, Ruhshdi (1991) shows that the introduction of a time trend variable in a translog total cost function may give rise to unsolvable problems of multicollinearity.

A major shortcoming of the translog functional form is that, since the natural logarithm of zero is not defined, it can only be used for multiproduct producers that supply positive quantities in all outputs. This problem can be solved by incorporating a Box-Cox transformation of the output variables. However, the translog functional form incorporating this transformation is non-linear in its parameters and therefore harder to estimate. In particular,it contains an increased number of parameters.[70]

In conclusion we may state that while the translog functional form has many desirable attributes, the applied economists should always keep in

---

70 See Greene (1993) for the econometric problems of the Box-Cox transformation and Caves et. al. (1981) for an application of this approach.

mind the objective of his research in his choice of the functional form. Relevant considerations are the quality and the quantity of available data, the computer resources available, and the multicollinearity problems that may arise.

7.3 Output characteristics

In specifying a cost model for a firm the applied production theory economist has to analyze carefully the production process of the firm and determines the inputs employed and the outputs produced.

The production reality shows that the majority of firms produce several outputs and that each output type has a different impact on costs. For instance, the production of 100 kWh of electricity during the peak period of electricity demand is more expensive than the production of 100 kWh during the off-peak period, because during the peak period of demand it is possible that less efficient and more costly electricity generators are employed. Further, two electric utilities that annually generate the same amount of electricity might have different production costs because of different system load curves or different service area sizes and structures.

During the past decades, economists have formulated a theoretical framework to analyze the cost structure of multiproduct firms.[71] Therefore, from a theoretical point of view, no problems arise by specifying a multiproduct cost function with 2, 3, or more outputs. Applied economists encounter more difficulties than theoretical economists when analyzing multiproduct cost functions for at least two reasons. First, data bases with detailed information on outputs and with sufficient degree of freedom to estimate a multiproduct cost function with, for example 15 outputs, are still not available. Second, output variables are usually highly correlated, exacerbating the problem of multicollinearity which, as presented in section 7.2, prevents a precise interpretation of parameter estimates. For these reasons applied economists have tried to develop alternative approaches to

---

[71] See Panzar and Willig (1977).

incorporate the heterogeneity of output in empirical analysis of cost functions.[72]

Generally, we can distinguish between three approaches. In the first approach, some outputs are aggregated in order to reduce their large number. To reduce the relatively large number of outputs, some outputs are aggregated while others are not. The second approach suggests the introduction of the cost model specification that includes several output characteristics variables, which should capture the heterogeneity dimension of the outputs, along with aggregated outputs. For example, the following cost function some output characteristics,

$$C = f(\mathbf{y}, \mathbf{w}, \mathbf{q}_{h+k}) \tag{7.10}$$

where C is total cost, **y** is a vector of outputs, **w** is a vector of input prices, and $\mathbf{q}_{h+k}$ is a vector of variables reflecting output characteristics. The load factor[73] and the size of the electric utility service area are examples of output characteristic variables for the electric industry.

The third approach is an extension of the second approach and suggests the estimation of a cost function employing panel data and using a fixed-effects model, which concerns itself with the extraction of firm specific effects by using dummy variables for different firms.[74] The fixed-effects model specification precludes biases in the coefficients of the included explanatory variables that might arise from the exclusion of unobserved output characteristics variables that are constant over time for a firm and are correlated with the other explanatory variables (*omitted variable bias*). For example, the fixed-effects specification for the cost function (7.10) can be expressed as:

$$C = f(\mathbf{y}, \mathbf{w}, \mathbf{q}_h, \mathbf{d}, \mathbf{t}) \tag{7.11}$$

---

72 See Oum and Thretheway (1989) for an interesting discussion of the heterogeneity of outputs and cost function specifications.

73 Load factor (LF) = kilowatt-hours during the year/ (8,760* maximum kilowatts during the year), where 8,760= 365 days * 24 hours.

74 The fixed-effects model, a term used in applied production analysis, is also called the Least Square Dummy Variables (LSDV) model.

where **y**,**w** as before, $\mathbf{q}_h$ is a vector of time variant output characteristic variable, **d** is a vector of indicator variables to reflect firm-specific effects and **t** a vector of time shifts. The coefficients of the indicator variables captures the firm-specific unobserved output characteristics assumed to be constant over time. Therefore, given availability of panel data and using a fixed-effects model, it is possible to control for differences in output characteristics that are specific to firms but are generally invariant over time by introducing firm-specific dummy-variables. As pointed out by Mundlak (1978), with the fixed-effects model specification, the estimates of the cost function parameters are free from bias due to correlation between included and excluded regressors to the extent that the excluded regressors are constant over time for a given firm.[75]

It is important to note that if we use the fixed-effects model there is no way we can estimate the coefficients of time-invariant ouput characteristics variables, because **d** captures the effect of all these variables.

The estimation of cost function using panel data and employing a fixed-effects model represents a valid approach to solve, albeit partially, the problem of output heterogeneity. This approach is employed in chapters 7 and 8, where the cost structure of the Swiss electric distribution utilities is analyzed.

7.4 Review of the literature

In this section we describe and evaluate some recent studies on the cost structure and the issue of scale economies in the electricity generation and distribution industry.[76] The cost structure of the electric utility industry has been studied extensively over the past twenty years, following the development of the duality theory which allows a technology to be analyzed using production, cost or profit functions. Since most contributions limit

---

75 For an application of this approach in estimating cost functions see for example Mundlak (1978), Caves, Christensen and Tretheway (1984b), Callan (1991), Good et al. (1992) and Friedlander et al. (1993).

76 For a extensive review of previous studies see Cowing and Smith (1978).

their attention to estimation of cost functions for the generation of electric power, little research is available on the cost structure including the transmission and distribution of electricity.[77]

The studies reviewed in this section are based on cost function models which offer an important application of duality theory in production analysis. Four studies on electricity generation and three studies on electricity distribution cost functions are reviewed. Table 7.2 and Table 7.3 summarize the main characteristics of each of the reviewed studies.

Nelson (1989)

Nelson analyzes the cost structure of the electric generation industry in the US by employing firm-level data for a cross-section of 22 firms over the period 1961-1983. The purpose of Nelson's study was to examine the question of scale economies and capacity utilization in the US electric generation industry. To evaluate these two issues a translog variable cost function was developed. The fixed-effects cost model specification employed by Nelson presents variable costs as a linear function of the following variables:

- output, measured in kWh
- capital stock, measured by the capital inventory method
- fuel and labor prices
- firm specific dummy variables
- a time trend to capture the effect of technological progress.

77 See Christensen and Greene (1976), Gollop and Roberts (1983), Nelson (1989), Callan (1991), Thompson and Wolf (1993) and Kamerschen and Thompson (1993) for examples of studies in electricity generation; See Weiss (1975), Henderson (1985), Roberts (1986), Nelson and Primeaux (1988) and Salvanes and Tjφota (1994) for empirical evidence for transmission and distribution of electricity.

Table 7.2 Review of previous studies on electricity generating cost functions

Study	***Data sample***	***Cross-section / Time series/ Panel data***	***Model used****	***Fixed effects model specification***	***Existence of economies of scale***
Nelson (1989)	*22 US electric utilities*	*panel data (1961-1983)*	$VC = f(y, k, w_f, w_l, d_i, T)$	*YES*	*YES*
Callan (1991)	*35 US electric utilities*	*panel data (1951-1984)*	$C = f(y, w_f, w_l, w_c, d_i, T)$ $VC = f(y, k, w_f, w_l, d_i, T)$	*YES*	*YES*
Rushdi (1991)	*1 Australian electric utility*	*time series data (1950-1984)*	$C = f(y, w_f, w_l, w_c, LF, T)$	*NO*	*YES; further, load factor has a negative effect on cost*
Thompson and Wolf (1993)	*130 US electric utilities*	*cross-section data (1985)*	$C = f(y, w_f, w_l, w_c)$	*NO*	*YES*

* *y = output measured in kWh; w_l, w_c, w_f are the prices of the labor, capital and fuel input respectively; d_i = firm specific dummy variable; k = capital stock; T = time trend variable and LF = load factor.*

Table 7.3 Review of previous studies on electricity distribution cost functions

Study	***Data sample***	***Cross-section / Time series/ Panel data***	***Model used****	***Fixed effects model specification***	***Existence of economies of density***	***Existence of economies of size***
Roberts (1986)	*65 US electric utilities*	*cross-section data (1978)*	$C = f(y_h, y_l, as, n, w_p, w_l, w_t, w_d, T)$	*NO*	*YES*	*NO*
Primeaux and Nelson (1988)	*23 US electric utilities*	*panel data (1961-1976)*	$C = f(y, as, n, tm, w_p, w_l, d_c, T)$	*NO*	*YES*	*NO*
Salvanes and Tjφota (1994)	*100 Norwegian electric distribution utilities*	*cross-section (1988)*	$VC = f(y, n, tm, w_p, w_l, k)$	*NO*	*YES*	*NO*

* *y = output measured in kWh; y_h = high voltage kWh; y_l = low voltage kWh; as = area size; tm = length of transmission lines; n = number of customers; $w_f, w_l, w_c, w_p, w_t, w_d$, are the prices of the fuel input, labor, capital, purchased power, transmission and distribution lines respectively; d_i = firm specific dummy variable; k = capital stock; T = time trend variable and LF = load factor.*

Nelson estimated jointly the variable cost function and the cost share equations for labor and fuel. The estimated cost function failed to satisfy the convexity condition with respect to fixed input. Further, Nelson reported substantial economies of scale and over-capacity for US electric generation utilities.

The major shortcoming of the Nelson study is that the employed translog variable cost function does not consider output characteristics that vary over time and that are important in explaining cost structure of electric utilities, such as the load factor. Therefore, the estimated values of scale economies and capacity utilization may be biased.

Callan (1991)

Callan studied electricity generation cost using data for a cross-section of 35 firms over the period 1951-1984. The objective of Callan's study was to analyze the productivity growth and the economies of scale in the US electric industry using three different approaches. The first approach is based on a methodology suggested by Solow (1957), while the other two approaches rely on the econometric estimation of a long- and short-run cost model. To estimate the total and the variable cost functions, Callan used the translog functional form and employed a fixed-effects model specification. The dependent variable of the long-run model is total cost while in the short-run model it is variable cost. The explanatory variables are one output, measured in kWh, three input prices, 21 firm-specific dummy variables and a time trend in the total cost function, whereas in the variable cost function the price for capital is replaced by capital stock. As opposed to the capital stock measure of Nelson, Callan used steam generating capacity. Moreover, he defined the price for capital as a firm's ratio of the capital expenditure to steam generating capacity.

The parameter estimates for the translog total and variable cost functions indicate the existence of economies of scale. Moreover, the estimation results of the total cost function satisfy all properties of a theoretically valid cost model, whereas the estimation results of the variable cost function failed to satisfy the convexity condition with respect to fixed input, with the implication that variable cost could be reduced by shedding capital. Callan interpreted this result as an indication of considerable overcapitalization of electric utilities. The only shorthcoming of this study is that the model specification does not consider the impact of intensity of use on the cost of a plant.

Rushdi (1991)

Rushdi took a somewhat different approach toward analyzing electricity generation costs than the two studies outlined above. Rushdi investigated the cost structure of the electric generation industry in Australia by employing time series data from 1950 to 1984 for one utility. He specified a translog total cost function with, as explanatory variables, the quantity of electricity generated, measured in kWh, the input prices for fuel, labor and capital, and the load factor. The interesting point of the model specification suggested by Rushdi is the introduction of the load factor as an explanatory variable in the cost function. With the inclusion of the load factor as a regressor, Rushdi was able to estimate a cost elasticity with respect to the load factor and thus to take into account the impact of intensive use on the cost of a plant.

Rushdi's results indicated substantial existence of scale economies in the Australian electricity generation industry. Moreover, the results showed that the load factor has a significantly negative impact on total cost. At the mean of the data, the cost elasticity with respect to the load factor was estimated to be -0.347 indicating that a 1% improvement in load factor will reduce cost by 0.35%. This result confirms the thesis suggested by electrical engineers that the costs to supply electricity extensively depend on the magnitude of the load factor.

While the Rushdi study can be commended for considering load factor effects, there are two shortcomings in its empirical results. One of these concerns is a possible loss of efficiency associated with its limited number of observations, 35, employed to estimate a translog cost function with 21 parameters. The second relates to the failure of its analysis of technological change. Rushdi attempted to estimate the translog total cost function with the inclusion of a time trend but, the multicollinearity problem that was discussed in section 7.2 appears in his data. In view of the small number of observations available and the many parameters to be estimated, this is not surprising.

Thompson and Wolf (1993)

The Thompson-Wolf study focused upon the question of regional differences in nuclear and fossil-fuel electricity generation costs. In contrast to the other studies introduced above, Thompson and Wolf presented the results of the estimation of a translog total cost function for the combined fossil-fuel electricity generation utilities and for the nuclear electricity generation ones. Moreover, separate translog total cost functions for regional samples of utilities were estimated.

To estimate these models, Thompson and Wolf employed a data set for the year 1985 on a cross-section of 90 US fossil-fuel electric utilities and a data set for the same year on a cross-section of 40 US nuclear electric utilities. Further, the combined fossil-fuel sample was divided into three regional samples.

The specified cost functions for fossil-fuel generation utilities and for nuclear power utilities are similar and simple. Total cost is regressed on produced kWh, on the prices of labor, capital and fuel.

The empirical results reported by Thompson and Wolf show that between nuclear and fossil-fuel generation utilities, important differences in costs and economies of scale exist. For both types of electric utilities the values of scale economies diminish as firm size increases. Moreover, the values of scale economies for the fossil-fuel utilities are smaller than those for nuclear utilities.

This study represents an interesting attempt to analyze the cost structure of different types of electric generation utilities. However, it failed to incorporate the heterogeneity of output in the cost model. For example, a load factor variable for which, as outlined by the authors, information is available, was not included in the model specification.

Roberts (1986)

Roberts's study of electricity supply was the first attempt to estimate a cost function for electricity production and distribution using the duality theory and a flexible functional form. Previous cost studies on electricity production and distribution focused on simpler relationships between cost, output and service area characteristics.

The main difference between the study of Roberts and the studies presented above is that Roberts analyzed the cost structure of vertically integrated electric utilities, whereas the studies of Nelson (1989), Callan (1991), Rushdi (1991) and Thompson and Wolf (1993) limited their analysis to the generation activities of electric utilities.

The specified cost model assumes that firms follow a two-stage production optimization process. In the first stage they make decisions regarding electricity generation, while in the second they make decisions concerning electricity distribution. Thus, in the second stage, firms are viewed as purchasing power either from their own plants or from other suppliers.

The data used by Roberts consists of a cross-section of 65 privately owned vertically integrated electric utilities in 1978. The electricity generation activity of these utilities was very high. Approximately 80 percent of kWh input was self-generated, while the remaining 20 percent was purchased from other electric generation utilities. Therefore, the Roberts study does not consider only firms which strictly distribute power to final consumers.

Roberts analyzed the cost structure of this vertically integrated industry by specifying the total production and delivery costs as the dependent variable, low- and high-voltage deliveries, the price of a kWh, the price of labor, and the price of transmission and distribution capital as explanatory variables. To control for other determinants of production and delivery costs, the square miles of services area and the total number of customers served were also included. Therefore, Roberts tried to capture the heterogeneous nature of kWh deliveries by including two outputs as well as two variables characterizing these to outputs as explanatory variables in the cost function. In this way, economies of density, customer density, and size may be distinguished. Economies of density, whose definition is close to the traditional definition of economies of scale, are defined as the proportional increase in total cost resulting from a proportional increase in outputs, holding all other explanatory variables fixed. Economies of customer density are measured as the proportional increase in total cost brought about by a proportional increase in outputs and the number of customers, holding all other explanatory variables fixed. Finally, economies of size are defined as the proportional increase in total cost brought about by a proportional increase in outputs, area size and number of customers.

The empirical results showed the existence of substantial economies of density and the existence of small economies of customers and economies of size in the production and distribution of electricity. Therefore, Roberts's study indicate that there is no strong evidence that larger service areas result in any economies of power delivery.

While this study offered one of the more complete analysis of the cost structure of vertically integrated utilities, it still suffers from three shortcomings. First, it failed to take into account the impact on total cost of the rate of output, e.g., the shape of the load system curve of a utility. Second, we suspect that the total number of customers included in the model specification as an explanatory variable is highly correlated with the capital stock of the electric utilities. Thus, this variable could be interpreted as a proxy for the capital stock, causing a problem in the model specification. Finally, the estimation results presented by Roberts might be biased because the output and output characteristics variables included in the model may be correlated with several output characteristics variables not included. As already presented, the use of a fixed-effects model, which requires a panel data set, could be a solution to this problem. However, the area size and the number of customers of the utilities are variables that generally are invariant over time, causing him to disappear in a fixed-effects cost model specification. This exclusion would solve the multicollinearity problem and the left-out variables problem, but has the disadvantage of preventing the measurement of cost elasticities with respect to these output characteristics. However, the empirical results of a fixed-effects cost model specification are unbiased, while those of a non fixed-effects specification might be biased and thus, not useful from a public policy point of view.

Nelson and Primeaux (1988)

Nelson and Primeaux specifically set out to test the impact of competition on production and distribution costs as well as the existence of economies of scale. Nelson and Primeaux analyzed the cost structure of a sample of 23 vertically integrated municipally owned electric utilities operating in the US during the period 1961-1976. However, while Roberts utilized a cross-section of data, Nelson and Primeaux employed panel data.

Thespecified cost model of Nelson and Primeaux explained total cost as a function of output, miles of transmission line, city size in squares miles, price of a kWh, wage rate, a time trend and a dummy variable for the nature of the competitive environment where the utility operates. Firms are divided into two groups, those operating in duopoly markets and those in monopoly markets.

Nelson and Primeaux defined two types of economies of scale, which correspond to the definition of economies of density and customer density employed by Roberts.

The empirical results indicate the existence of economies of density and the lack of existence of economies of customer density. Moreover, the introduction of competition into a monopoly market reduces the average production and distribution costs.

The results of the study of Nelson and Primeaux have to be interpreted carefully for two reasons. First, as admitted in the paper by the authors, the variable referred to the mile of transmission line is actually a proxy for the level of capital stock, which implies that the specified cost model lies between a short- and a long-run cost model. In fact, to be a true short-run cost function specification, the dependent variable of the cost function should be variable cost and not total cost as defined by Nelson and Primeaux. Therefore, results presented cannot be interpreted in the of the authors. We suspect that the estimation results obtained in the study suffer from the same econometric problems described in the presentation of Roberts' study results.

Salvanes and Tjφota (1994)

Salvanes and Tjφota's study of electricity supply is the first attempt to estimate a cost function only for distribution companies using duality theory and a flexible functional form.

The main difference between the study of Salvanes and Tjφota and the studies on the electricity distribution cost function presented above is that Salvanes and Tjφota analyzed the cost structure of electric distribution utilities, whereas the studies of Roberts (1986) and Nelson and Primeaux (1988) employed a sample of vertically integrated companies.

The data used by Salvanes and Tjϕota consists of a cross-section of 100 Norwegian public electricity distribution utilities in 1988. These companies behave as local natural monopolies, whose activities include transportation of electricity to end users, installation and maintenance of equipment and administration. Therefore, the Salvanes and Tjϕota' s study considers only firms that distribute power to final consumers.

Salvanes and Tjϕota analyzed the cost structure of the Norwegian electricity distribution industry by specifying the delivery variable cost as the dependent variable and the total voltage delivered, the price of a kWh, the price of labor and the capital stock, measured by the length of the transmission lines, as explanatory variables. To control for other determinants of delivery cost, the total number of customers served were also included. As Roberts (1986) and Nelson and Primeaux (1988), Salvanes and Tjϕota distinguish between economies of density and economies of size. However, the definition of economies of size adopted by Salvanes and Tjϕota corresponds to the definition of economies of customer density suggested by Roberts (1986). Therefore, the results with regard to economies of scale, density and size reported in Tables 6.2 and 6.3, have to be made carefully.

The results reported in this study indicate that there are economies of density but only moderate economies of size (in terms of Roberts's study, moderate economies of customer density). Moreover, as in the study of Callan (1991) and Nelson (1989) the coefficient of the capital stock variable is found to be positive, which implies the violation of the non-increasing regularity condition at the median of the data.

After this brief discussion on the results and methods of seven recent cost function studies, we would like to present several guidelines, which we believe to be important for the applied economist when specifying and estimating a cost function for electric utilities:

- the cost function should take into account the multi-dimensional nature of output. For example, peak output does not have the same impact on production and distribution costs as off-peak output. If data on peak and off-peak output are not available, then the inclusion in the cost function of the load factor can partially solve this data problem;

- the cost function should be estimated using a fixed-effects model specification and therefore, using panel data. According to Mundlak (1978), this approach overcomes the bias of the estimation results that can occur in the presence of unobserved firm's effects that are correlated with the regressors;
- the estimation of a translog short-run cost function requires particular attention be paid to potential multicollinearity problems that can arise between the output and the capital stock variable;
- to develop a better description of the cost structure of electric it is best to estimate both a short- and long-run cost model and then to compare the results.

In the next two chapters we present a specification and estimation of a short- and long-run cost function for the Swiss electricity distribution utilities. Two translog cost functions are estimated using a fixed-effects model to control for differences in output characteristics and operating environments that are specific to electric utilities but are generally invariant over time. Moreover, the load factor is included in the cost models as an explanatory variable.

As presented before, the estimation of a cost function requires two assumptions: firms should face exogenous outputs and input prices and firms are assumed to adjust input levels to minimize costs.

The first assumption is easy to justify for the Swiss electricity distribution companies. First, these companies are regulated local monopolies required under an "obligation to serve" to supply all the electricity demanded at the prices approved by the regulatory commission. Second, as all electricity distribution utilities employ essentially the same technology, they compete with each other in electricity, labor and specialized capital equipment purchases. As there are thousands of Swiss electricity distribution companies, there is little potential for the existence of monopsony power in these input markets.

The second assumption is supported by the following two reasons. First, these local monopolies have incentives to minimize costs because they can earn an unconstrained profit. This assumption is supported by the fact that the majority of these companies effectively earn a profit. Second, in a spatial sense, the shareholders of these companies generally are communities and local investors. Therefore, they are knowledgeable about the local market and its technological context, and they can provide utilities with incentives to minimize costs.

CHAPTER 8

THE LONG-RUN COST MODEL

In the previous chapter we presented the economic theory of producer behavior and some general aspects of applied production analysis. This chapter introduces an analysis of the cost structure of the Swiss electricity distribution industry based on the theoretical and empirical frameworks presented in the previous chapter.

In what follows, the specification and estimation of a translog total cost function for electric distribution utilities are discussed first. The estimation results are then presented and used to analyze scale economies. Section five provides a more intuitive interpretation of scale economies with a graphical presentation of the average and marginal total cost curves. Moreover, a detailed analysis of the impact of load factor on total cost is introduced. A brief discussion concludes the chapter.

8.1 Model specification

The electricity utilities analyzed in this study operate within an urban area and are involved in generation, transmission and distribution of electricity. However, the amount of generated power is small and is determined by the ability to exploit favorable hydroelectric power generation possibilities. These electric distribution utilities purchase power from ten large vertically integrated electric utilities and then distribute it to their residential, industrial and commercial customers.

Inputs to the electricity distribution process consist primarily of labor, capital and electricity. Capital stock takes the form of electricity distribution lines, transformers, meters and, in a small part, electricity generators.

The electricity distribution costs comprise electricity purchase, labor and capital costs. Labor costs are primarily incurred in customer service, facility maintenance and general administration. Capital costs are incurred in the depreciation of existing capital stock, new transformers or transmission lines and plant retirements.

The review of the literature discussed in chapter 7 revealed that the majority of the studies on cost structure of electric utilities adopted a single-product cost function specification. However, because of the multi-dimensional nature of output in electricity generation and distribution, output is not completely defined by kWh. Previous researchers were well aware of this fact, as Stewart (1979) indicates when he states that the nature of output must be thought of in at least two dimensions, i.e., a volume to be supplied over the year and an instantaneous rate, which measures potential. Electricity is a non-storable good and a good with cyclic demand that varies among daily hours. Thus, because electricity has to be produced when it is demanded, two different cyclic electricity consumption patterns imply different instantaneous rates, different production and distribution capacities and, therefore, different production costs. Figure 8.1 shows two daily system load curves, which are obtained by plotting on a time axis the total electricity consumption at a given hour.

In spite of a numerical similarity, distributing 1,000,000 kWh in a city with a daily system load curve characterized by peak and off-peak periods of electricity demand as shown in Figure 8.1 (a), is more expensive than distributing 1,000,000 kWh in a city with a load curve similar to that shown in Figure 8.1 (b). As pointed out by Parson (1939), the costs to supply electricity depends largely on the magnitude of the load factor.

In this study, to capture the multi-dimensional nature of output, we introduced in the cost function the load factor as an output characteristic variable, along with output measure themselves, kWh. Moreover, using pooled time-series and cross section data it is possible to employ a fixed-effects model which reduces bias in the estimated cost function parameters arising from unmeasurable output characteristics that are constant over time for a given electric utility and correlated with the other explanatory variables.[78]

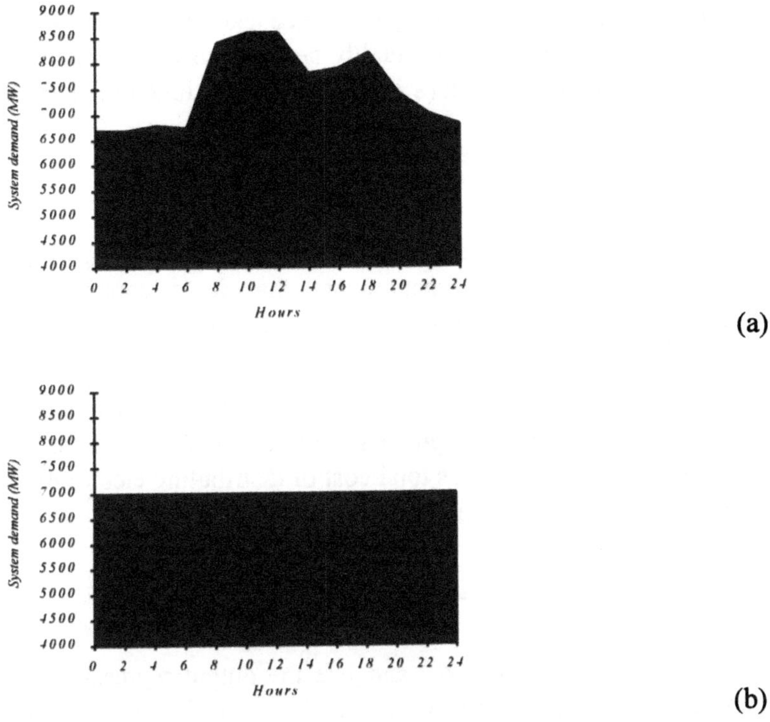

Figure 8.1 Daily system load curves

When an electric utilitiy produces one output, we can represent its production function by

$$y = F(x_1, x_2, \ldots\ldots, x_g; q_1, q_2, \ldots. q_n; T) \tag{8.1}$$

78 See section 7.3 of this study.

where y denotes output, x_j (j = 1,......, g) denotes inputs, q_e (e = 1,....., n) denotes operating and output characteristics variables, and T is a vector of time shifts representing an index of technology.

As long as the production function satisfies the usual regularity conditions, the electricity utility is a price taker in input markets, and a cost minimizer, then the theory of duality between cost and production ensures that the production function (8.1) has a dual cost function of the form

$$C = C(y; w_1, w_2, \ldots\ldots, w_g; q_1, q_2, \ldots\ldots q_n; T) \tag{8.2}$$

where w_j (j = 1,......, g) denotes input price corresponding to input x_j. The properties of the cost function are that it is concave and linearly homogeneous in input prices and non-decreasing in input prices and output.[79]

Output is measured by the total number of kWh delivered. Inputs to the electricity distribution process consist primarily of labor, capital and purchased power. The firm's total cost of distributing electricity can then be represented by the cost function

$$C = C(y, w_c, w_p, w_l, LF, T, FD) \tag{8.3}$$

where C represents total cost and y is the output represented by the total number of kWh delivered. w_c, w_p and w_l are the prices of capital, kWh input and labor, respectively. LF is the load factor introduced in the model as output characteristic. The introduction of the load factor in the model should capture the impact on cost of the intensity of use of the plant.[80] T is a time variable which captures the shift in the technology representing change in technical efficiency and FD is a vector of firm-specific binary variables included to control for firm fixed effects. These firm effects can be given the interpretation of omitted variables that are specific to the firm and display little variability over the sample period considered. Therefore, the estimation

79 See Cornes (1992), p.106.

80 See Foreman-Peck and Waterson (1985) for a discussion of the introduction of load factor in cost models.

procedure involves a fixed-effects technique, taking advantage of the panel of data that is available. In our model we view the time invariant fixed-effects as unobserved/unmeasured firm specific heterogeneity like the output characteristics, the characteristics of the service area or the aspects of a firm's management.[81]

Estimation of cost function (8.3) requires the specification of a functional form. As pointed out in chapter 7, the translog cost function offers an appropriate functional form for answering questions about economies of scale.[82] Most important for our purposes, it imposes no a priori restrictions on the nature of technology, allowing the values for scale economies and substitution elasticities to vary with output. The translog approximation to (8.3) is

$$\begin{aligned} ln(\frac{C}{w_p}) = {} & \alpha_0 + \eta_y \, ln\, y + \sum_j \beta_j \, ln(\frac{w_j}{w_p}) + \phi_{LF} \, ln\, LF + \frac{1}{2}\eta_{yy}(ln\, y)^2 \\ & + \frac{1}{2}\gamma_{LFLF}(ln\, LF)^2 + \frac{1}{2}\sum_j \sum_i \mu_{ji} \, ln(\frac{w_j}{w_p}) ln(\frac{w_i}{w_p}) \\ & + \sum_j \omega_{yj} \, ln\, y \, ln(\frac{w_j}{w_p}) + \sum_j \theta_{jLF} \, ln(\frac{w_j}{w_p}) ln\, LF \\ & + \rho_{yLF} \, ln\, y \, ln\, LF + \sum_f \kappa_f DF_f + \tau_T T \end{aligned} \tag{8.4}$$

where $j = c, l$ and $f = 1,2,...,n$. T, the time trend is included in a way to capture the effects of neutral technical change. Note that by normalizing total cost and input prices by one of the input prices, we impose the theoretical condition that the cost function is linearly homogeneous in input prices.

---

81 For a similar attempt to capture firm specific differences in the electric industry that remain unchanged over the period see Nelson (1989) and Callan (1991).

82 A translog function requires the approximation of the underlying cost function to be made at a local point, which in our case, is taken at the median point of all variables. Thus, all independent variables are normalized at their median point.

In order to improve the efficiency of the estimation of least squares parameter estimates for the cost function, a cost system is estimated. This system consists of the translog cost function (8.4) and the factor share equations. By applying Shephard's lemma, the resulting share equations take the familiar form:

$$S_j = \beta_j + \sum_j \mu_{ji} \ln(\frac{w_i}{w_p}) + \omega_{yj} \ln y + \theta_{jLF} \ln LF \tag{8.5}$$

where S_j is the share of input j in total costs. The price by which we normalize is that of the input whose share equation we dropped from the estimating system.

One advantage of the translog function, as well as any other flexible form, is that a variety of assumptions regarding the technology can be tested. In this study the following restrictions on the technology are tested: homotheticity and a Cobb-Douglas technology. Homotheticity of the cost function implies that the ratio of any two factor demand functions is independent of the output level and that the elasticity of total cost with respect to output is independent of factor prices. Whereas a Cobb-Douglas technology is characterized by unitary elasticities of substitution and a constant elasticity of cost with respect to output, i.e., it is the same for all values of the output.

In addition to these technological restrictions, the hypothesis that the fixed effects are identical for all electricity distribution utilities is tested. This simply requires replacing the $\kappa_f DF_f$ with a single regression intercept.

Table 8.1 presents the parametric restrictions imposed on the translog cost function (8.4) to test these hypothesis. All tests are conducted by estimating the set of equations (8.4) and (8.5) with and without appropriate restrictions imposed.

Table 8.1 Translog total cost restrictions for tests of alternative technology hypothesis and identical fixed effects

Hypothesis	Parameter restrictions
• Homotheticity	$\omega_{yj} = 0$
• Cobb-Douglas	$\omega_{yj} = 0$, $\eta_{yy} = 0$, $\gamma_{LFLF} = 0$, $\mu_{ji} = 0$, $\theta_{jLF} = 0$ and $\rho_{yLF} = 0$
• Identical fixed effects	$\kappa_f = 0$

The appropriate test for these hypotheses is the likelihood ratio test, where the test statistic, $-2 \ln (L_R / L_U)$, is asymptotically distributed as chi-squared with r degrees of freedom, where L_R is the maximum value of the likelihood function for the restricted model and L_U is the corresponding value for the unrestricted model and r is the number of independent restrictions imposed. The hypothesis is rejected if the calculated value exceeds the critical value.

8.2 Data and estimation procedure

The models are estimated for cross-sectional samples of publicly owned electricity distribution utilities operating in Swiss cities. In Switzerland, there are approximately 130 companies that could be included in a study of city electricity utilities. The Swiss Federal Statistical Office and the Swiss Federal Energy Office collect financial data only for a sample of about 60 utilities serving cities. Of the companies listed in this sample, 21 are not appropriate for the purpose of our analysis because the amount of self-generated electricity is high. Since the aim of this study is to analyze the cost structure of distribution, companies that had more than 20 percent of their capital invested in generating activities in 1989, were excluded.

Placing the above restrictions on the available data results in a sample of 39 city electricity distribution utilities for which appropriate data are available. For estimation, panel data for four years 1988, 1989, 1990 and

1991 have been used. The primary data sources were the Swiss Federal Statistical Office's *Wertschöpfungsstatistik* and the Swiss Federal Energy Office's *Finanzstatistik*; additional data were taken from the annual publication of the Swiss Cities Association. Data on the number of employees, maximum peak demand and distributing capacity were obtained from a questionnaire sent to the utilities.

The necessary data include the total costs, the shares in total costs of labor, capital and electricity purchased costs, the load factor, the prices of purchased power, capital and labor as well as the quantity of kWh delivered. All input prices, total cost and variable cost were deflated to 1982 constant Swiss francs using the Consumer Price Index.

Total cost is simple total expenditure as annually reported by the companies and consists of the annual capital cost, labor cost and electricity purchased cost. Average yearly wage rates are estimated as labor expenditures, which consist of the sum of wages, and employee pensions and benefits, divided by the number of employees. This implicitly assumes that all employees in each category work the same number of hours and, on average, have the same quality. The average price of purchased power is constructed according to the approach used by Nelson and Primeaux (1988) by dividing the expenditure on purchased power by the kWh deliveries. The capital price is the most difficult variable to measure because capital is purchased in one time period but delivers a flow of services over subsequent time periods. From a theoretical point of view the best approach to measure capital costs and capital price is to apply the methodology proposed by Jorgenson (1967, 1974) which is based on the measurement of capital stock by the capital inventory method and on a definition of capital price based on the relationship between the acquisition price of new capital goods and the present value of all future services received. Unfortunately, no data is available which would allow the application of the approach proposed by Jorgenson (1967, 1974). Therefore, following Friedlander and Wang Chang (1983) and Filippini and Maggi (1993) the capital price is calculated from the residual capital costs divided by the capital stock. Residual costs is total costs minus labor and energy costs. According to Callan (1992) and Nelson (1990), the capital stock is defined as total installed distributing capacity.

Table 8.2 contains each variable's first quartile, median and third quartile. There are clearly considerable variations across utilities in all variables.

Table 8.2 Descriptive statistics

Variables	Unit of measurement	25.Percentile	Median	75.Percentile
Total cost	SwF.	9,116,800	13,491,000	38,078,000
Labor price	SwF. for worker unit	51034	64726	76382
Purchased power price	SwF./kWh	0.07	0.081	0.089
Capital price	SwF. for a unit of capital	46.929	65.057	90.110
Load factor		0.50	0.53	0.57
Output	kWh	73,350,000	99,500,000	307,350,000

The total cost function (8.4) and the cost share equations (8.5) are the set of equations to be estimated. Additive disturbance terms were appended to (8.4) and each of the share equations (8.5). By including the cost share equations, additional degrees of freedom are gained without the need to estimate any additional coefficients. We assume that the resulting disturbance vector is independently and identically multivariate normally distributed with mean 0 and a constant non-singular covariance matrix. The system was estimated using the iterative Zellner's technique (Zellner, 1962) for seemingly unrelated regressions (SUR). This allows for the possibility that there may be a correlation between the disturbance terms of the cost and share equations for each observation, but that such correlation does not exist across observations within a given equation. As the share equations sum to one, an equation must be omitted from the system to implement SUR. However, using the iterative Zellner technique, the resulting estimates are equivalent to maximum likelihood estimates (Kmenta and Gilbert, 1974), and they are invariant to which share equation is deleted (Barten, 1969).

8.3 Estimation results

In this section we report the econometric results obtained from estimating the total cost function model specified in equation (8.4) and (8.5) and using the data described in section 8.2. We also present the results obtained from estimating three others versions of the cost function model. The first, corresponds to a homothetic production technology. The second version represents a Cobb-Douglas cost function, while the third version of the cost function, which we refer to as the "pooled data" version, restricts the fixed effects so that they are the same for all electric utilities.

The estimated coefficients and their associated standard errors of these four versions of the cost function model are presented in Table 8.3. The estimated functions are well behaved. Most of the parameter estimates are statistically significant and there is a medium degree of similarity between the coefficients obtained from the four versions of the total cost function. As expected, the estimation results of models 1 and 2 are analogous, whereas these of models 3 and 4, in terms of load factor and output coefficients are different. However, the estimation results of model 4 may be biased because of excluded variables such as the characteristics of the service area that are constant over time for a given utility and are correlated with the regressors.

Since total cost and the regressors are in logarithms and have been normalized, the first order coefficients are interpretable as cost elasticities evaluated at the sample median. All these coefficients have the expected signs and are highly significant. The output elasticity is positive and implyies that an increase in the production of output will increase total cost. A 1% increase in the delivery of power will increase the total cost in models 1 and 2 by approximately 0.82% and in the models 3 and 4 by 0.64% and 0.95%, respectively.

Table 8.3 Total cost parameter estimates (standard errors in parentheses)

Coefficient	Model 1[a]	Model 2[b]	Model 3[c]	Model 4[d]
α_0	16.646 ***	16.634 ***	16.738 ***	16.414 ***
	(0.018)	(0.018)	(0.017)	(0.007)
η_y	0.823 ***	0.839 ***	0.646 ***	0.949 ***
	(0.022)	(0.022)	(0.019)	(0.005)
β_l	0.126 ***	0.129 ***	0.133 ***	0.117 ***
	(0.003)	(0.002)	(0.002)	(0.003)
β_c	0.279 ***	0.274 ***	0.221 ***	0.285 ***
	(0.003)	(0.002)	(0.005)	(0.003)
ϕ_{LF}	-0.204 ***	-0.217 ***	-0.045 **	-0.364 ***
	(0.029)	(0.030)	(0.019)	(0.038)
η_{yy}	-0.019	-0.009	-	-0.016 ***
	(0.019)	(0.019)		(0.002)
μ_{cc}	0.166 ***	0.161 ***	-	0.143 ***
	(0.004)	(0.004)		(0.004)
μ_{ll}	0.026 ***	0.029 ***	-	0.050 ***
	(0.005)	(0.005)		(0.003)
γ_{LFLF}	0.206 ***	0.224 ***	-	0.259 ***
	(0.057)	(0.058)		(0.058)
ω_{yl}	0.006 **	-	-	-0.012 ***
	(0.003)			(0.002)
ω_{yc}	-0.011 ***	-	-	-0.019 ***
	(0.003)			(0.002)
ρ_{yLF}	0.013	0.002	-	0.046 ***
	(0.020)	(0.020)		(0.010)
θ_{cLF}	-0.166	-0.172	-	-0.155
	(0.017)	(0.017)		(0.016)
θ_{lLF}	-0.061 ***	-0.063 ***	-	-0.024
	(0.016)	(0.016)		(0.015)
μ_{cl}	0.009 ***	0.008 ***	-	0.015 ***
	(0.003)	(0.003)		(0.003)
τ_T	0.001	0.001	-	-0.008 ***
	(0.007)	(0.007)		(0.001)
Generalized R^2 for the system	0.997	0.997	0.982	0.988
Log Likelihood	1094.30	1090	771.36	910.93

*, **, ***: significantly different from zero at the 90%, 95%, 99% confidence level.

a. unrestricted fixed-effects model

b. homothetic version of the fixed-effects model

c. Cobb-Douglas version of the fixed-effects model

d. pooled data model

The cost elasticity with respect to load factor is negative in all versions of the cost model, indicating that a 1% improvement in load factor will reduce cost by approximately 0.20% in models 1 and 2, by 0.05% and 0.36% in models 3 and 4, respectively.

The labor, purchased power and capital cost shares are positive, implying that the cost function is monotonically increasing in input prices. In all versions of the cost model the purchased power accounts for approximately 61% of the electricity distribution utilities costs while labor accounts for approximately 12% and capital for the remaining 27% of total cost. Moreover, in models 1, 2 and 3, approximately 90% of the fixed-effects coefficients (firm-specific dummy variables) were significantly different from 0.[83]

Parameter estimates of the four versions of the translog cost function satisfy the regularity condition of concavity in input prices at the median point of approximation, which requires that the own-price elasticities of inputs be negative and that the Hessian Matrix, $\left[\partial^2 C / \partial w_i \partial w_j\right]$, be negative semi-definite. Because homogeneity in input prices and symmetry of the second order terms were imposed, the estimated functions satisfy all regularity conditions of a theoretically valid total cost model.

To determine which of the four versions of the cost function model is statistically preferable we employ the Likelihood Ratio test presented before by maintaining the unrestrictive model 1 as full specification and treating models 2, 3 and 4 as restricted models.

Table 8.4 shows the results of this test. With the aid of critical values of the chi-square statistics given in Table 8.4, the set of restrictions on the parameters of the translog cost function implied by homotheticity, Cobb-Douglas technology, identical fixed-effects can be rejected at the five percent significance level. Model 1 is, therefore, used in all the following analysis.

---

83 Tables 8.3 omits the estimated coefficients of the firm specific dummy variables; a copy of these coefficients is presented in Appendix B1.

Table 8.4 Likelihood Ratio test for parametric restrictions[a]

Hypothesis	Likelihood Ratio	Critical value (0.05 level)
• Homotheticity	8.60	5.99
• Cobb-Douglas technology	183.37	19.67
• Identical fixed-effects	645.88	55.75

a. All tests are relative to model 1.

8.4 Estimated economies of scale

Returns to scale are usually defined in terms of the relative increase in output resulting from a proportional increase in all inputs. Therefore, returns to scale are defined along an arbitrary input-mix ray. However, Hanoch (1975) noted that it is more relevant to measure returns to scale by the relationship between total cost and output along the expansion path. Returns to scale and economies of scale will differ unless the production function is homothetic. Since the results of the Likelihood Ratio test presented in Table 8.4 rejects the homotheticity hypothesis we follow Hanoch's concept of economies of scale. Scale economies exist if increasing production lowers average cost. Following Christensen and Greene (1976) and Caves et. al. (1984b) we define an extent of economies of scale (ES_C) as the proportional increase in total costs brought about by a proportional increase in output. This is equivalent to the inverse of the elasticities of total cost with respect to output,

$$ES_C = \frac{1}{\dfrac{\partial \ln C}{\partial \ln y}} \tag{8.6}$$

which may be expressed in terms of the coefficients of (8.4) as

$$ES_C = \frac{1}{\eta_y + \eta_{yy} \ln y + \sum_j \omega_{yj} \ln(\frac{w_j}{w_p}) + \rho_{yLF} \ln LF} \quad (8.7)$$

We will talk of economies of scale if ES_C is greater than 1 and, accordingly, identify diseconomies of scale if ES_C is below 1. In the case of $ES_C = 1$, no economies or diseconomies of scale exist.

Table 8.5 Overall scale economy estimates

Utilities size	small y = 73,350,000 kWh	medium y = 99,500,000 kWh	large y = 307,350,000 kWh
ES_C (load factor = 0.3)	1.22	1.23	1.26
ES_C (load factor = 0.5)	1.21	1.22	1.25
ES_C (load factor = 0.8)	1.20	1.21	1.24

The empirical evidence strongly supports the hypothesis of scale economies in the electricity distribution industry. Table 8.5 presents the estimates of scale economies evaluated with the estimation results from model 1. In order to gain a better idea of economies of scale in this industry, we evaluated equation (8.7) for small, medium, and large electricity utilities, respectively.[84] Moreover, three different sets of economies of scale, evaluated at three levels of the load factor, are presented.

The results indicate that when small, medium and large companies simultaneously increase the production of output, total cost will increase less than proportionally. Furthermore, the importance of scale economies increases with size implies the existence of a somewhat "L-shaped" long-

84 The economies of scale have been evaluated at the input prices of the median company.

run average cost curve.[85] Moreover, the results for the three different levels of the load factor show that the lower the load factor the higher the value of the economies of scale.

Table 8.6 Estimates of scale economies for 5 small and 5 large electric distribution utilities in 1991

Size of the electric distribution utility	*Estimate of scale economies*
small	1.21
small	1.19
small	1.20
small	1.21
small	1.20
large	1.29
large	1.28
large	1.29
large	1.30
large	1.32

An estimate of scale economies can be calculated for individual electric distribution utilities by evaluating equation (8.7) at the observed level of output, load factor and input prices. We selected the 5 utilities producing the smallest amount of output and the 5 utilities producing the largest amount of output. The smaller utilities are those with less than 55,000,000 kWh, whereas the larger utilities are those with more than 650,000,000 kWh. Due to the given confidentiality of our data, we cannot present descriptive statistics for each electric distribution utility. The

85 "L-shaped " long-run average cost curves are frequently observed in the real world. See Koutsoyiannis (1979) for a discussion on the shape of costs.

estimates of scale economies for these selected utilities in 1991 are presented in Table 8.6.

The range of estimates of scale economies in our sample is from 1.20 to 1.33. This means that all the electric distribution utilities in our sample are operating in the downward sloping portion of their actual average cost function.

8.5 Marginal and average total costs

Another way to illustrate the extent of scale economies is to examine average and marginal costs. The relation among the extent of scale economies, average total cost (AC) and marginal total cost (MC) is

$$ES_C = \frac{1}{\frac{\partial \ln C}{\partial \ln y}} = \frac{1}{\frac{\partial C}{\partial y}\frac{y}{C}} = \frac{1}{\frac{MC}{AC}} \tag{8.8}$$

Table 8.7 presents the values of the average and marginal total costs for small, medium and large electric distribution utilities calculated at various levels of the load factor.[86] These values have been evaluated by using the following formulas which are derived from (8.8)

$$MC = \frac{C}{Y}(\eta_y + \eta_{yy}\ln y + \sum_j \omega_{yj}\ln(\frac{w_j}{w_p}) + \rho_{yLF}\ln LF) \tag{8.9}$$

$$AC = \frac{e^{\left(\begin{array}{l}\alpha_0 + \eta_y \ln y + \sum_j \beta_j \ln(\frac{w_j}{w_p}) + \phi_{LF}\ln LF + \frac{1}{2}\eta_{yy}(\ln y)^2 + \frac{1}{2}\gamma_{LFLF}(\ln LF)^2 + \frac{1}{2}\sum_j\sum_i \mu_{ji}\ln(\frac{w_j}{w_p})\ln(\frac{w_{ij}}{w_p}) \\ + \sum_j \omega_{yj}\ln y \ln(\frac{w_j}{w_p}) + \sum_j \theta_{jLF}\ln(\frac{w_j}{w_p})\ln LF + \rho_{yLF}\ln y \ln LF + \sum_f \kappa_f DF_f + \tau_T T\end{array}\right)}}{Y} \tag{8.10}$$

---

[86] The average and marginal total costs have been evaluated at the input prices of the median company.

Table 8.7 Marginal and average total costs for small, medium and large utilities with different load factor

Utilities size	small y = 73,350,000 kWh	medium y = 99,500,000 kWh	large y = 307,350,000 kWh
load factor = 0.3			
Marginal total cost	0.18	0.17	0.13
Average total cost	0.21	0.20	0.16
load factor = 0.5			
Marginal total cost	0.15	0.14	0.10
Average total cost	0.18	0.17	0.14
load factor = 0.8			
Marginal total cost	0.14	0.13	0.10
Average total cost	0.17	0.16	0.13

Marginal total cost ranges from Swf. 0.10 to Swf. 0.18, while average total cost ranges from Swf. 0.13 to Swf. 0.21. Both are decreasing in output, which implies "L-shaped" average and marginal total cost curves. Further, electric distribution utilities characterized by low load factors show higher average and marginal costs than utilities with high load factors.

The results on scale economies, average and marginal costs can be conveniently summarized by illustrating the average and marginal total cost curves.

Figure 8.2 shows these curves over a relevant range of output, with other exogenous variables held constant at their median values. The figure clearly illustrates the pattern of scale economies and average and marginal total costs discussed above.

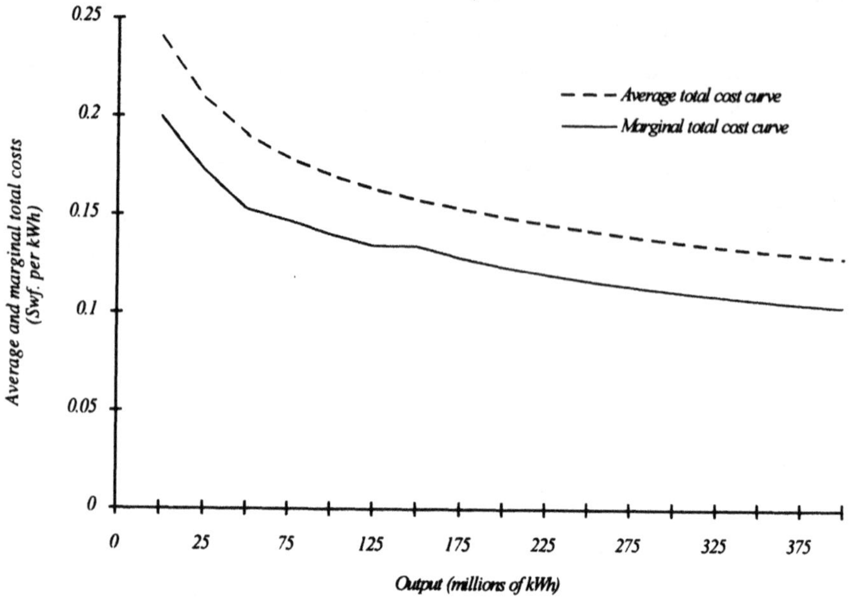

Figure 8.2 Average and marginal total cost curves

8.6 Cost effects of load factor

The fact that electric distribution utilities face cyclic demand, which implies an intensity of distributing capacity utilization that varies according to hour and season, raises the question of the effect of the load factor on total costs. An electric distribution utility which serves customers with strong daily and seasonal variations in demand would be expected to have higher costs than one serving customers with little daily and seasonal variations in demand. Moreover, the knowledge of the effect on total cost of the load factor allows us to verify the efficacy of time-of-use rates as an instrument to promote efficient utilization of electric distribution utilities.

In studies on the cost structure of electric generation utilities, Stewart (1979) and Rushdi (1991) came to the conclusion that improvement of the load factor can significantly reduce costs.To analyze the impact of the shape of the load curve on total costs we evaluated the cost elasticity with respect to load factor for the 5 small and 5 large utilities presented above. This cost elasticity is given by

$$CE_{LF} = \frac{\partial \ln C}{\partial \ln LF} = \phi_{LF} + \gamma_{LFLF} \ln LF + \sum_j \theta_{jLF} \ln(\frac{w_j}{w_p}) + \rho_{yLF} \ln y \qquad (8.11)$$

The values of cost elasticity with respect to load factor for the 5 smaller and 5 larger utilities of our sample are presented in Table 8.8.

Table 8.8 Cost elasticity with respect to load factor for small and large utilities in 1991

Size of the electric distribution utility	Estimate cost elasticity
small	-0.29
small	-0.22
small	-0.29
small	-0.30
small	-0.34
large	-0.36
large	-0.26
large	-0.34
large	-0.31
large	-0.26

The results also show that in the case of the Swiss electricity distribution utilities an improvement in load factor can significantly reduce costs. This finding is highly consistent with prior expectations. At higher levels of load factor, electric distribution utilities can be expected to require smaller capacity and require smaller amounts of labor for maintanance, employee supervision and personnel related to overhead activities. Further, the difference between the values of cost elasticities of small electric utilities and large ones is not evident. Thus, we can conclude that large utilities have the same possibilities to reduce costs by improving the load factor as the small ones.

8.7 Elasticities of factor demand and substitution

In this section we will briefly analyze the properties of the input demands which can be expressed with respect to the prices.[87] The own-price (OPE_{ii}) and the cross-price (CPE_{ji}) elasticities of input demand are defined to be

$$
\begin{aligned}
OPE_{ii} &= \frac{\delta \ln x_i}{\delta \ln w_i} && \text{for all } i \\
CPE_{ji} &= \frac{\delta \ln x_j}{\delta \ln w_i} && \text{for all } j \neq i
\end{aligned}
\tag{8.12}
$$

where x_i is the firm's derived demand for input i. The Allen Elasticities of Substitution (AES_{ji}) in term of cost function parameters and factor shares (S_j) are defined as

87 See Chambers (1988) for an interesting discussion of the elasticities of factor demand and substitution.

$$AES_{ii} = \frac{\dfrac{\delta^2 \ln C}{\delta(\ln w_i)^2} + S_i^2 - S_i}{S_i^2} \qquad \text{for all } i$$

(8.13)

$$AES_{ji} = \frac{\dfrac{\delta^2 \ln C}{\delta \ln w_j \ \delta \ln w_i} + S_j S_i}{S_j S_i} \qquad \text{for all } j \neq i$$

For the translog total cost function (8.4) these elasticities are computed using the following formulas:[88]

$$CPE_{ji} = (\mu_{ji} + S_j S_i)/S_i \qquad \text{for all } j \neq i$$
$$OPE_{ii} = (\mu_{ii} + S_i S_i - S_i)/S_i \qquad \text{for all } i \qquad (8.14)$$

$$AES_{ji} = (\mu_{ji} + S_j S_i)/S_i S_j \qquad \text{for all } j \neq i$$
$$AES_{ii} = (\mu_{ii} + S_i S_i - S_i)/S_i^2 \qquad \text{for all } i \qquad (8.15)$$

Tables 8.9 and 8.10 present the own (OPE_{ii}) and cross-price (CPE_{ji}) input elasticities and Allen elasticities of substitution (AES_{ji}) evaluated at the median values of the data. As shown in Table 8.9 the estimated own price elasticities of input demand have the correct sign. When the own-prices rise, purchases of inputs decline. Further, all input demands are inelastic. This suggests that electric distribution utilities have more flexibility in adjusting labor in response to higher labor prices than capital or electricity in response to higher capital or electricity prices, respectively.

---

88 See Binswanger (1974).

Table 8.9 Own and cross-price elasticities of factor demand

	Labor	Capital	Electricity
Labor	-0.67	0.16	0.07
Capital	0.16	-0.13	-0.01
Electricity	0.32	-0.03	-0.08

To analyze the substitution possibilities between different pairs of inputs in the production technology we concentrate our attention on the Allen elasticities of substitution presented in Table 8.10.

Table 8.10 Allen elasticities of substitution

	Labor	Capital	Electricity
Labor	-5.30	1.26	0.53
Capital		-0.45	-0.05
Electricity			-0.14

Positive elasticities show that labor and capital are substitutes, as are labor and electricity. The negative sign of the elasticity of substitution between capital and electricity indicate small complementary relationships in the technology.

8.8 Concluding comments

The estimation results of the long-run cost function indicate the existence of scale economies in the Swiss electricity distribution industry. The value of the estimated scale economies for small, medium, and large companies range between 1.20 and 1.26. These findings are consistent with the existence of a somewhat "L-shaped" long-run average cost curve, which indicates the electricity distribution industry as a natural monopoly. In addition, the estimation results show that improvement in load factor can significantly reduce costs.

Finally, it has to be observed that all the results presented in this chapter are based on the estimation of a long-run total cost function, which carries the assumption that all inputs, including capital, are employed at their cost-minimizing levels. Given the nature of capital in the electric power industry this assumption could be violated, resulting in imprecise estimates of economies of scale, average and marginal costs and input demand elasticities. In order to verify this assumption, therefore, in the next chapter the results of the estimation of a short-run variable cost function are presented.

CHAPTER 9

THE SHORT-RUN COST MODEL

The fact that the capital employed in the production and distribution of electricity is long-lived and that electricity is a good characterized by cyclical demand and non-storability supports the thesis that these utilities might not have used the optimal level of capacity through the period considered by this analysis. Thus, this chapter presents the specification and estimation of a variable cost function in which the capital stock is considered a quasi-fixed input[89]. The addressed issues presented in this chapter include conditions for long-run cost minimization, a detailed analysis of economies of scale and utilization for different electric distribution utilities, a graphical presentation of the marginal and average variable costs and short-run input demand elasticities.

9.1 Model specification

Most of the studies on the cost structure of the electricity distribution industry have utilized long-run cost functions which invoke the assumption that electricity distribution utilities are in static equilibrium, using all inputs at their optimal levels. This chapter contends that this assumption is fallacious and presents a specification that allows for the possibility that firms are not in static equilibrium with respect to one factor

---

89 This chapter is based on a paper published in the *Applied Economics*. See Filippini (1996).

of production, the stock of capital. If it is the case that the utilities are not in equilibrium with respect to this quasi-fixed input, then measures of economies of scale based on estimates of the long-run cost function may be inprecise.

There are two plausible arguments supporting the claim that electricity distribution utilities have not maintained the optimal level of capacity throughout the period of our study. First, the capital embodied in the distribution lines and transformers is long-lived, thus adjustment to a change of the time profile of electricity demand is costly. Furthermore, distribution capacity is planned and built on the basis of long-term load forecasts which can be highly inaccurate. Second, municipalities give an exclusive territorial franchise to a distribution utility in exchange for guaranteed service to all resident consumers. Therefore, the distribution utilities are legally obliged to maintain excess capacity so as to meet sudden increases in demand and to guarantee service. This again argues for a quasi-fixed capital stock in electricity distribution.

The restricted variable cost function is used to model the production structure of the electricity distribution utilities. This variable cost function takes account of divergence from the optimum in that the quantity of physical capital cannot be adjusted to achieve minimum total cost during the period of observation for a given set of input prices and the quantity of output. With the distribution utility producing one output, we can represent its production function by

$$y = F(x_1, x_2, \ldots\ldots, x_g; k_1, k_2, \ldots\ldots, k_c; q_1, q_2, \ldots\ldots, q_n; T) \tag{9.1}$$

where y denotes output, x_j (j = 1,......, g) denote inputs, k_u (u = 1,.....,c) denote quasi-fixed inputs, q_e (e = 1,....., n) denotes operating and output characteristics variables, and T is a vector of time shifts representing an index of technology.

If the transformation function satisfies certain regularity conditions (Lau 1976), and if utilities minimize variable costs, there exists a variable cost function of the form

$$C = VC(y; w_1, w_2, \ldots\ldots, w_g; k_1, k_2, \ldots\ldots, k_c; q_1, q_2, \ldots\ldots, q_n; T) \tag{9.2}$$

where w_j (j = 1,......, g) denotes prices of inputs x_j. The properties of the variable cost function (9.2) are that it is concave and linearly homogeneous in input prices, non- decreasing in input prices and outputs, and decreasing in the quasi-fixed inputs.[90]

Inputs to the electricity distribution process consist primarily of labor, purchased power and quasi-fixed input capital. Therefore, the variable cost function of a Swiss electricity distribution utility may be written as

$$C = VC(y, w_p, w_l, k, T, LF, FD_i) \tag{9.3}$$

where VC represents variable cost and y is the output represented by the total number of kWh delivered. w_p and w_l are the prices of the kWh input and labor, respectively. k is the stock of capital and LF is the load factor, which should capture the impact on cost of the intensity of use of the plant. However, the impact of load factor on variable cost should be lower than that on the total cost as described in chapter 8, because the shape of the load curve generally influences the fixed costs more than the variable costs. T is a time variable which captures the shift in the technology representing change in technical efficiency and FD is a vector of firm-specific binary variables included to control for firm fixed effects. As already pointed out in chapter 8, we view the time invariant fixed effects as unobserved/unmeasured firm specific heterogeneity like the output characteristics, the characteristics of the service area or the aspects of the management of the firm.

As in chapter 8, we specify the variable cost function in a translog form. The translog approximation to (9.3) is

---

90 See Chambers (1988), p.102.

$$ln(\frac{VC}{w_p}) = \alpha_0 + \eta_y \, ln\, y + \beta_l \, ln(\frac{w_l}{w_p}) + \kappa_k \, ln\, k + \gamma_{LF} \, ln\, LF + \frac{1}{2}\eta_{yy}(ln\, y)^2$$

$$+ \frac{1}{2}\mu_{ll}(ln(\frac{w_l}{w_p}))^2 + \frac{1}{2}\kappa_{kk}(ln\, k)^2 + \frac{1}{2}\gamma_{LFLF}(ln\, LF)^2$$

$$+ \omega_{yl} \, ln\, y \, ln(\frac{w_l}{w_p}) + \varsigma_{ky} \, ln\, y \, ln\, k + \rho_{yLF} \, ln\, y \, ln\, LF + \pi_{lk} \, ln(\frac{w_l}{w_p}) ln\, k \tag{9.4}$$

$$+ \delta_{lLF} \, ln(\frac{w_l}{w_p}) ln\, LF + \phi_{kLF} \, ln\, k \, ln\, LF + \sum_f \kappa_f DF_f + \tau_T T$$

where $f = 1,2,...,n$. T, the time trend is included so that it captures the effects of neutral technical change. The theoretical condition that the cost function is homogeneous of degree one in input prices is imposed by normalizing variable cost and labor price by w_p.

Degrees of freedom can be saved by estimating the firm's cost share equations together with its variable cost function. By applying Shephard's Lemma to (9.4) we get the cost share equation for labor input as

$$S_l = \beta_l + \mu_{ll} \ln(\frac{w_l}{w_p}) + \omega_{yl} \ln y + \pi_{lk} \ln k + \delta_{lLF} \ln LF \tag{9.5}$$

As in the case of the total cost function, we would like to use the estimated coefficients of the variable cost function to test statistically the hypotheses of homotheticity, Cobb-Douglas technology and identical fixed-effects. Table 9.1 shows the restrictions imposed on the translog variable cost function (9.4) to test these restrictions. The test statistic used is the Likelihood Ratio test.

Table 9.1 Translog variable cost restrictions for tests of alternative technology hypothesis and identical fixed effects

Hypothesis	Parameter restrictions
• *Homotheticity*	$\omega_{yl} = 0$
• *Cobb-Douglas*	$\omega_{yl} = 0$, $\eta_{yy} = 0$, $\gamma_{LFLF} = 0$, $\kappa_{kk} = 0$, $\mu_{ll} = 0$
	$\zeta_{ky} = 0$, $\pi_{lk} = 0$, $\delta_{lLF} = 0$, $\phi_{kLF} = 0$ *and* $\rho_{yLF} = 0$
• *Identical fixed effects*	$\kappa_f = 0$

9.2 Data and estimation procedure

The translog variable cost function (9.4) is estimated for the same sample of publicly owned electricity distribution utilities operating in Swiss cities described in chapter 8. The necessary data for the estimation of the variable cost function (9.4) are the same as these for the estimation of the total cost function (8.4) with the following exceptions:

- variable costs, defined as the sum of labor and expenditure on purchased power, replaces total costs as the dependent variable, and variable input cost shares replace the respective total input cost shares.
- the price of capital is replaced with a measure of the capital stock.

Generally, two different measures of capital stock can be defined: a capacity measure and a cost measure.[91] For example, the total installed distributing capacity and the length of the distribution lines are capacity

---

[91] For a generasl discussion of the measurements of the capital inputs and capital price see Jorgenson (1963, 1967), Christensen and Jorgenson (1969) and Bruse and Fuhrmann (1981).

measures.[92] Whereas the capital stock of a company computed using the perpetual inventory method or, more directly, computed as the book value of equipment, transformer, distribution lines and land, are cost measures.

Using the perpetual inventory method the capital stock is calculated as

$$k_t = \frac{\sum NI_t}{PI_t} \qquad (9.6)$$

where t is the time period , NI is net investment, PI is the price index and k is the capital stock. Net investment is found by subtracting net capital stock in time t from net capital stock in time t-1. The number of time periods chosen is the life of the asset. Moreover, to account for the heterogenous nature of capital goods, the capital stock can be calculated separately for different assets and then, using for example a Tornquist approximation to a Divisia index, form indices of capital stocks.

The capital stock calculated using the perpetual inventory method and assuming a one-hoss depreciation pattern is the most interesting measure of capital stock because first, it is based on economic costs, second it considers that in the electricity distribution industry the flow of services received from capital is constant until its retirement, third, it allows one to take into account heterogeneity of capital goods, such as differences between equipment and plant, and differences in the vintage and efficiency of these capital goods. Unfortunately, the information contained in our data set are not sufficient to calculate it.

---

92 Generally, these two measures are highly correlated. For instance, in our data set these two variables show a coefficient of correlation of 0.9.

Table 9.2 Descriptive statistics

Variables	*Unit of measurement*	*25.Percentile*	*Median*	*75.Percentile*
Variable cost	*SwF.*	6,689,000	10,159,000	23,772,000
Price of labor	*SwF. per worker*	51,034	64,726	76,382
Price of purchased power	*SwF./kWh*	0.07	0.081	0.089
Capital	*Installed distributing capacity inKW*	43,283,000	62,500,000	137,520,000
Load factor		0.50	0.53	0.57
Output	*kWh*	73,350,000	99,500,000	307,350,000

The book value of equipment, transformer, distribution lines and land is not an interesting measure of capital because it is based on accounting costs; moreover, it does not take into account that the flow of services received from capital is constant until its retirement.

The total installed distributing capacity and the length of the transmission lines have the disadvantage of being engineering measures rather than economic measures of capital stock. Further, these proxy variables fail to consider the heterogeneity of capital goods. However, its advantage is to take into account that the flow of services received from capital is constant at full productive efficiency until its retirement. Due to lack of data to calculate the capital stock using the perpetual inventory method, the total installed distributing capacity was chosen as an indicator of capital stock.[93]

Descriptive statistics of the variables are presented for small, medium and large electricity distributing firms in Table 9.2. All input prices and variable costs were deflated to 1982 constant Swiss francs using the Consumer Price Index.

The system of equations (9.4) and (9.5) was estimated employing Zellner's (1962) iterative seemingly unrelated estimation technique.

93 See Callan (1991) and Nelson (1990) for the same approach in defining a proxy for the capital stock.

9.3 Estimation results

Table 9.3 presents the coefficient estimates of the variable cost model, together with their standard errors under four cases. Model 1 represents estimates of the variable cost function model specified in equations (9.4) and (9.5). In model 2 we assume homotheticity in output. Model 3 presents the cost function estimates corresponding to a Cobb-Douglas technology. Whereas model 4, the "pooled data" model, restricts the fixed effects so that they are the same for all electric utilities.

Table 9.3 shows that the majority of the coefficients of the four versions of the variable cost model are statistically significant and, unlike to the results obtained in the estimation of the total cost model, many estimates differ among models 1, 2, 3 and model 4, including the output, the capital stock and load factor term. However, as pointed out in the previous chapter, the estimation results of model 4 may be seriously biased, because this model specification does not account for excluded variables that are constant over time for a given utility and are correlated with the regressors.

A well defined variable cost function should be increasing with respect to output and input prices, concave with respect to input prices and non-increasing with respect to capital stock.

The coefficients of output and input prices have the expected signs and are highly significant in all versions of the variable cost model. The cost elasticity of output is less than one in models 1, 2 and 3 and equals 1 in model 4. This measure indicates, for example, that a 1% increase in electric power deliveries will increase distribution variable costs by approximately 0.86% in models 1, 2 and 3 and by 1% in model 4.

The labor cost share are positive, implying that the variable cost function is monotonically increasing in input prices. As might be expected, the proportion of purchased power costs in the variable cost model is very high with a value of 88%. Further, we find that the condition of concavity in input prices, which implies that the Hessian matrix, $\left[\partial^2 C / \partial w_i \partial w_j\right]$, be negative semi-definite, is satisfied at the median of the data in all versions of the model.

Table 9.3 Variable cost parameter estimates (standard errors in parentheses)

Coefficient	Model 1[a]	Model 2[b]	Model 3[c]	Model 4[d]
α_0	16.281*** (0.016)	16.279 *** (0.016)	16.278 *** (0.015)	16.100 *** (0.007)
η_y	0.853 *** (0.019)	0.859 *** (0.019)	0.871 *** (0.016)	1.014 *** (0.022)
β_l	0.128 *** (0.003)	0.127 *** (0.003)	0.129 *** (0.002)	0.123 *** (0.003)
κ_k	0.037 *** (0.016)	0.036 ** (0.016)	0.013 (0.008)	0.014 (0.022)
γ_{LF}	0.020 (0.020)	0.016 (0.019)	-0.011 (0.015)	-0.374 *** (0.044)
η_{yy}	0.053 (0.054)	0.051 (0.055)	-	-0.127 *** (0.046)
κ_{kk}	0.071 (0.046)	0.082 * (0.045)	-	-0.075 * (0.044)
μ_{ll}	0.030 *** (0.004)	0.029 *** (0.004)	-	0.055 *** (0.002)
γ_{LFLF}	0.067 (0.081)	0.084 (0.080)	-	0.125 (0.157)
ω_{yl}	-0.013 (0.010)	-	-	-0.005 (0.009)
ρ_{yLF}	-0.021 (0.054)	-0.028 (0.054)	-	0.305 *** (0.070)
ζ_{ky}	-0.069 (0.048)	-0.075 (0.048)	-	0.099 ** (0.044)
δ_{lLF}	-0.040 *** (0.017)	-0.050 *** (0.015)	-	-0.113 *** (0.019)
π_{lk}	0.024 ** (0.010)	0.011 *** (0.003)	-	0.003 (0.009)
ϕ_{kLF}	0.021 (0.050)	0.034 (0.049)	-	-0.354 *** (0.079)
τ_T	0.000 (0.000)	0.000 (0.000)	-	-0.002 ** (0.001)
Generalized R^2 for the system	0.999	0.999	0.999	0.993
Log Likelihood	856.22	855.54	817.40	633.43

*, **, ***: significantly different from zero at the 90%, 95%, 99% confidence level.

a. unrestricted fixed-effects model
b. homothetic version of the fixed-effects model
c. Cobb-Douglas version of the fixed-effects model
d. pooled data model

The coefficient of capital stock is in all models positive, pointing to increases in variable costs with increases in capacity at the sample median. This result indicates that the regularity condition of non-increasing variable cost with respect to the capital stock is not satisfied at the median of the data. A more detailed discussion of the implications of a positive capital stock coefficient for the interpretation of the estimation results is given in the next section.

The coefficient of the load factor variable is not significantly different from 0 in models 1, 2 and 3, while significantely different from 0 in model 4. This result is not surprising because the load factor affects the total cost more than the variable cost.[94] Finally, in models 1, 2 and 3, approximately 95% of the fixed-effects coefficients (firm-specific dummy variables) are significantly different from 0. [95]

Table 9.4 presents the calculated values of the Likelihood Ratio statistics for the tests of assumptions regarding technology and identical fixed-effects presented in Table 9.1. As in the case of the total cost function presented in chapter 8, the results of this tests reject the Cobb-Douglas and identical fixed-effects hypotheses. However, the homotheticity hypothesis cannot be rejected at the 5% level of significance.[96] Therefore, the estimation of economies of scale, average and marginal costs, and price and substitution elasticities presented in the following sections are based on the estimation results of the homothetic fixed-effects model.

---

94 In the previous chapter we found a significant negative influence of the load factor on total cost.

95 Tables 9.3 omits the estimated coefficients of the firm specific dummy variables; a copy of these coefficients is presented in Appendix B2.

96 The homotheticity concept in a variable cost function differs from that in a total cost function because of the inclusion in the variable cost function of quasi fixed inputs.

Table 9.4 Likelihood Ratio test

Hypothesis	Likelihood Ratio	Critical value (0.05 level)
• Homotheticity	1.36	3.84
• Cobb-Douglas	77.64	19.67
• Identical fixed effects	445.58	55.75

9.4 Long-run equilibrium conditions

As presented in section 7.1, a total cost function can be obtained by first minimizing a variable cost function and then minimizing total costs with respect to quasi fixed-inputs, by holding the variable inputs at their short-run level. To determine the corresponding long-run cost function with this two-step minimization approach, the short-run variable cost function must be well behaved, i.e., increasing with respect to output and input prices, concave with respect to input prices, non-increasing with respect to quasi-fixed inputs and, additionally, convex in quasi-fixed inputs.[97] Therefore, if the translog variable cost function (9.4) satisfies these conditions, then is possible to infer long-run characteristics of the technology using the estimation results of (9.4) and the prices of the capital stock. In the case of the variable cost function (9.3), the first-order condition for long-run cost minimization is

$$\left.\frac{\partial VC}{\partial k}\right|_{k=k^*} + w_k = 0 \tag{9.7}$$

where w_k is the user cost of capital and k^* denotes the optimal value of capital stock which minimizes short-run total cost (SRC).

---

97 See Chambers (1988).

Equation (9.7) implies that utilities substitute capital for variable inputs until the marginal reduction in variable cost equals the user cost of capital. Therefore, one can interpret the derivative of VC with respect to the quasi-fixed factor as its shadow price. Moreover, the non-increasing regularity condition of the variable cost function implies that the shadow price approaches zero as the quantity of the quasi-fixed factor becomes unreasonably large.

Following Kulatilaka (1985) and Chambers (1988), if the electric utility employs a non-optimal level of the capital stock, the envelope condition (9.7) is violated, which for a given output level is interpreted as over- or under-utilization of the capital stock. The following rules can be applied

$$\textit{If} \quad \left.\frac{\partial VC}{\partial k}\right|_{k=k^*} \prec -w_k \quad \text{then } k \prec k^* \quad (\text{under} - \text{capitalization});$$

$$\textit{if} \quad \left.\frac{\partial VC}{\partial k}\right|_{k=k^*} = -w_k \quad \text{then } k = k^* \quad (\text{optimal} - \text{capitalization})$$

$$\textit{if} \quad \left.\frac{\partial VC}{\partial k}\right|_{k=k^*} \succ -w_k \quad \text{then } k \succ k^* \quad (\text{over} - \text{capitalization}).$$

With under-capitalization investment incentives exist, while with over-capitalization the electric utilities should disinvest.

In our case, the first order derivative of variable cost function with respect to capital stock (k) is given by

$$\frac{\partial VC}{\partial k} = \frac{VC}{k}\left(\kappa_k + \kappa_{kk} \ln k + \pi_{lk} \ln\left(\frac{w_l}{w_p}\right) + \varsigma_{ky} \ln y + \phi_{kLF} \ln LF\right) \qquad (9.8)$$

The empirical results cited in Table 9.3 indicate a positive derivative of the variable cost function with respect to the capital stock at the median of the data, which implies violation of the non-increasing regularity condition. Moreover, as presented in Table 9.5, the majority of smaller and larger electric distribution utilities of our sample also present a value of the derivative of variable cost function with respect to capital stock that violates the non-increasing regularity condition. This result, which confirms the results of numerable studies on the cost structure of electric power firms estimating a variable cost function,[98] indicates that marginal increases in capital stock cause an increase in variable costs rather than a decrease as normally expected in cost theory.

Table 9.5 Derivative of variable cost function with respect to capital stock for small and large utilities in 1991

Size of the electric distribution utility	*Value of the derivative of variable cost function with respect to capital stock*
small	3.73
small	1.25
small	-0.40
small	3.63
small	-0.45
large	4.21
large	7.16
large	4.32
large	5.41
large	5.29

In the literature we find two possible interpretations of this theoretically implausible sign of the coefficient of capital stock. The first interpretation, proposed by Cowing and Holtmann (1983) and largely employed in electricity cost studies literature,[99] argues that the positive sign of the coefficient of capital stock is an indicator of an excessive amount of

98 See for example Nelson (1989), Callan (1991), Hammond (1992) and Salvanes and Tjφota (1994).

99 See Callan (1991), Hammond (1992), Salvanes and Tjφota (1994).

capital stock employed by the firms. In this case, an increase of the capital stock would lead to an increase of both variable and fixed costs.

The second interpretation, proposed by Guyomard and Vermersch (1989) and sustained by Filippini (1991), supports the idea that the incorrect sign of the coefficient of the capital stock is probably derived from multicollinearity between the output and the capital stock. Therefore, for Guyomard and Vermersch (1989) and Filippini (1991) the existence of this econometric problem does not allow a precise interpretation of the sign of the capital stock.

We think that the interpretation of Cowing and Holtman (1983) is not completely satisfactory for two reasons. First, it is not supported by the economic theory of producer behavior, because it indicates a negative marginal product of capital stock. Second, Cowing and Holtman (1981) do not consider the econometric implications for the interpretation of the estimation results of the use of proxy variables for capital stock that are highly correlated with the output, whereas the interpretation suggested by Guyomard and Vermersch (1989) and Filippini (1991) is more precise because it is based on an empirical analysis of the problem. However, Guyomard and Vermersch (1989) and Filippini (1991) failed to state the causes of this empirical problem.

Generally, we believe that the main cause at the origin of the often found positive sign of the coefficient of capital stock in estimating variable cost functions for the electric industry has to be imputed to the empirical difficulty to define and calculate the capital stock variable. Especially the definition of the capital stock variable in a industry faced with peak load capacity requirements can give rise to empirical problems. Due to lack of data the majority of the short-run cost studies in the electric sector have used physical measures of the capital stock, such as generating capacity, distributing capacity or length of transmission lines of an electric utility. These proxy-variables for the capital stock reflect maximum available production capacity at one particular point in time and, therefore, are generally highly correlated with output increasing; hence, the possibility of a multicollinearity problem. Thus, the interpretation of the coefficient of the capital stock variable when employing a proxy variable for the capital stock which reflects the production capacity is not straightforward and cannot be generalized because of the potentially high correlation of this variable with the output.

In the case of the estimation results reported in Table 9.4, we believe that the positive sign of the capital stock coefficient has to be attributed to the employed proxy variable for the capital stock, which is a production capacity measure.

To determine whether multicollinearity between output and capital stock variables is causing a problem, we follow Judge et al. (1988) and Greene (1993) and examine the simple correlation coefficients among all the explanatory variables as well as the condition number of the matrix of these variables.

Table 9.6 presents the simple correlation coefficients between the explanatory variables of the translog variable cost function (9.4). A value of the simple correlation coefficient greater than 0.9 can be interpreted as a sign that a multicollinearity problem exists.[100] In Table 9.6 we find six values of the simple correlation coefficient greater than 0.9. As expected, these high correlation coefficients are observed between capital stock and output, and between the interaction variables that include output and capital.

These results are confirmed by the more sophisticated measure of multicollinearity proposed by Besley et al. (1980), the condition number, which is defined as the square root of the ratio of the largest to the smallest characteristic root of a matrix of variables. Besley et al. (1980) suggest calculating this ratio for the matrix, **X'X**, where **X** is the matrix of the explanatory variables in the variable cost model. The calculated condition number for the matrix of explanatory variables of the variable cost model (9.4) is 380, which indicated a potential multicollinearity problem.[101]

To solve this multicollinearity problem we calculated the condition number for the matrix of the explanatory variables susbstituting the distributing capacity variable with two other possible proxy variables for capital stock, the length of transmission lines and the book value of the firm. Unfortunately, with these two other proxy variables for the capital stock the

---

100 See Judge et al. (1988) for a presentation of measures for the severity of multicollinearity.

101 Belsley et al. (1980) suggest that a value of the condition number greater than 20 indicates a potential multicollinearity problem.

calculated index numbers also indicated a potential multicollinearity problem. We think that the more appropriate solution to this problem would be to calculate the capital stock variable in a more precise way by using the capital inventory method. The use of this method may change significantly the estimation results. Unfortunately, our data set do not allow the application of this method.

We can conclude this discussion on the possible explanations of the positive sign of the capital stock coefficient by saying that the estimation results presented in Table 9.3 show failure of the translog variable cost function to be non-increasing and convex in the capital stock at the median of the data. Further, the interpretation of these results suggested by Cowing and Holtman (1983) is not satisfactory. Therefore, these results are inconclusive regarding the long-run cost minimization hypothesis.

Table 9.6 Simple correlation coefficients (all variables are measured in natural logarithm)

	y	k	LF	w_l/w_p	y^2	k^2	LF^2	$(w_l/w_p)^2$	yk	yLF	yw_l/w_p	kLF	kw_l/w_p	w_l/w_p LF
y	1.00	0.95	0.08	0.34	0.85	0.78	0.21	-0.11	0.82	0.19	0.34	0.17	0.26	0.09
k		1.00	-0.06	0.30	0.83	0.77	0.08	-0.15	0.81	0.08	0.29	0.10	0.21	0.00
LF			1.00	0.02	0.09	-0.00	0.63	-0.08	0.02	0.73	0.06	0.56	-0.02	0.32
w_l/w_p				1.00	0.16	0.10	0.07	-0.04	0.13	0.05	0.24	0.01	0.01	-0.09
y^2					1.00	0.96	0.11	-0.15	0.98	0.22	0.37	0.29	0.34	0.07
k^2						1.00	-0.03	-0.14	0.99	0.11	0.32	0.23	0.35	0.02
LF^2							1.00	-0.10	0.00	0.61	0.06	0.28	-0.01	0.46
$(w_l/w_p)^2$								1.00	-0.15	-0.09	0.62	-0.04	0.61	-0.24
yk									1.00	0.14	0.34	0.25	0.34	0.03
yLF										1.00	0.11	0.9	0.04	0.50
yw_l/w_p											1.00	0.14	0.89	-0.24
kLF												1.00	0.12	0.42
kw_l/w_p													1.00	-0.24
$w_l/w_p\,LF$														1.00

9.5 Economies of utilization and scale

A measure of utilization economies is defined by Caves and Christensen (1988) as unity divided by a proportional increase in variable cost resulting from a proportional increase in output holding capital stock constant. Thus, economies of utilization represent variable cost changes when output is increased with capacity constant. According to this definition, we define economies of utilization (EU_{VC}) as

$$EU_{VC} = \frac{1}{\dfrac{\partial \ln VC}{\partial \ln y}} \tag{9.9}$$

which may be expressed in terms of the coefficients of (9.4) as

$$EU_{VC} = \frac{1}{\eta_y + \eta_{yy} \ln y + \omega_{yl} \ln(\dfrac{w_l}{w_p}) + \varsigma_{ky} \ln k + \rho_{yLF} \ln LF} \tag{9.10}$$

We will speak of economies of utilization if EU_{VC} is greater than 1 and, accordingly, identify diseconomies of utilization if EU_{VC} is below 1. In the case of $EU_{VC} = 1$ no economies or diseconomies of utilization exist.

Similarly, scale economies exist if increasing production lowers average cost. Following Caves et al. (1981) we define a measure of economies of scale as the proportional increase in total cost resulting from a proportional increase in output. The degree of the economies of scale (ES_{VC}) in the variable cost function can be measured by the formula,

$$ES_{VC} = \frac{1 - \dfrac{\partial \ln VC}{\partial \ln k}}{\dfrac{\partial \ln VC}{\partial \ln y}} \tag{9.11}$$

which may be expressed in terms of the coefficients of (9.4) as

$$ES_{VC} = \frac{1-(\kappa_k + \kappa_{kk}\ln k + \varsigma_{ky}\ln y + \pi_{lk}\ln(\frac{w_l}{w_p}) + \phi_{kLF}\ln LF)}{\eta_y + \eta_{yy}\ln y + \omega_{yl}\ln(\frac{w_l}{w_p}) + \varsigma_{ky}\ln k + \rho_{yLF}\ln LF} \qquad (9.12)$$

We will speak of economies of scale if ES_{VC} is greater than 1, and accordingly, identify diseconomies of scale if ES_{VC} is below 1. In the case of $ES_{VC} = 1$ no economies or diseconomies of scale exist.

The value of the economies of scale is a function of capital, as well as input prices, load factor and output. The procedure for measuring economies of scale with the formula (9.12) in the presence of quasi-fixed input have been continually discussed in the literature.[102] Caves et al. (1981) have proposed evaluating (9.12) at the actual capital stock, k, while Friedlander and Spady (1981) and Oum et al. (1991) suggested the evaluation of (9.12) at the equilibrium stock of capital, k^*. Using the actual capital stock, k, we measure economies of scale at the actual point of operation, whereas using the equilibrium stock of capital, k^*, we measure economies of scale along the expansion path where total cost is minimized at every level of output.

We decided to evaluate (9.12) at the actual capital stock because, as already presented in the previous section, the estimation results of the variable cost function reported in Table 9.3 do not allow us to determine the corresponding long-run cost function.

The measures of utilization and scale economies obtained from the estimates of the variable cost function along with the scale economies measures presented in the preceding chapter and calculated with the empirical estimates of the total-cost function (ES_C) are shown in Table 9.7. Further, the results for small, medium, and large electricity utilities are

---

102 See Caves et al. (1981), Friedlander and Spady (1981) Braeutigam and Daughety (1983), Nelson (1985) and Oum et al. (1991).

presented.[103] Note that all the economies measures reported in Table 9.7 indicate the presence of utilization and scale economies. Therefore, if small, medium and large companies increase output with holding capacity fixed, variable cost will increase less than proportionally. Moreover, an increase in output without holding capacity fixed will increase total cost less than proportionally. Furthermore, the importance of utilization and scale economies increases with size.

The empirical results in Table 9.7 indicate that the values of scale economies measured from a total-cost function are higher in magnitude than those estimated from the variable-cost function. If the industry under study has not been in long-run equilibrium, the estimation of scale economies from the coefficients of a variable cost function is more precise than that obtained from the coefficients of a total cost function. However, because of the multicollinearity problem presented in section 9.4, the values of scale and utilization economies calculated with the estimates of the translog variable cost function (9.4) have to be interpreted carefully.

Table 9.7 Economies of utilization and scale

Size of utilities	small	medium	large
	y = 73,350,000 kWh	y = 99,500,000 kWh	y = 307,350,000 kWh
EU_{VC}	1.18	1.16	1.09
ES_{VC}	1.11	1.12	1.15
ES_C	1.21	1.22	1.25

It is noteworthy necessary to observe that if a theoretically expected negative sign of the capital stock coefficient in (9.12) is introduced, the values of economies of scale would be higher than those calculated with the

103 The economies of scale and utilization have been evaluated at the input prices and load factor of the median company.

positive coefficient of capital reported in Table 9.7.[104] Nevertheless, the empirical results of the variable cost function confirm that Swiss electric distribution utilities are subject to economies of scale.

Table 9.8 presents the estimates of utilization and scale economies for the 5 smallest and 5 largest electric distribution utilities. The range of estimates of utilization and scale economies in our sample is from 1.11 to 1.25 and from 1.11 and 1.20, respectively. Theoretically, the value of scale economies should be higher than that of utilization economies. Therefore, the results reported in Table 9.8 disagree with the theory of producer behavior. Again, this discrepancy has to be imputed to the positive value of the capital stock coefficient.

Table 9.8 Estimates of utilization and scale economies for 5 small and 5 large electric distribution utilities in 1991

Size of the electric distribution utility	*Estimate of utilization economies*	*Estimate of scale economies*
small	1.15	1.11
small	1.11	1.11
small	1.11	1.12
small	1.15	1.12
small	1.11	1.11
large	1.20	1.17
large	1.21	1.17
large	1.20	1.17
large	1.24	1.19
large	1.25	1.20

104 A negative sign of the capital coefficient would increase the numerator of the formula (9.12) and thus increase the value of the economies of scale.

9.6 Marginal and average variable costs

From the translog variable cost function (9.4) it is possible to derive the formulas for marginal and average variable costs. The marginal (MVC) and average(AVC) variable costs are defined as

$$MVC = \frac{VC}{y}(\eta_y + \eta_{yy}\ln y + \omega_{yl}\ln(\frac{w_l}{w_p}) + \varsigma_{ky}\ln k + \rho_{yLF}\ln LF) \quad (9.13)$$

$$AVC = \frac{e^{\left(\begin{array}{l}\alpha_0 + \eta_y \ln y + \beta_l \ln(\frac{w_l}{w_p}) + \kappa_k \ln k + \gamma_{LF}\ln LF + \frac{1}{2}\eta_{yy}(\ln y)^2 + \frac{1}{2}\mu_{ll}(\ln(\frac{w_l}{w_p}))^2 + \frac{1}{2}\kappa_{kk}(\ln k)^2 \\ + \frac{1}{2}\gamma_{LFLF}(\ln LF)^2 + \omega_{yl}\ln y \ln(\frac{w_l}{w_p}) + \varsigma_{ky}\ln y \ln k + \rho_{yLF}\ln y \ln LF + \pi_{lk}\ln(\frac{w_l}{w_p})\ln k + \delta_{lLF}\ln(\frac{w_l}{w_p})\ln LF \\ + \phi_{kLF}\ln k \ln LF + \sum_f \kappa_f DF_f + \tau_T T\end{array}\right)}}{y} \quad (9.14)$$

Table 9.9 shows the values of the average and marginal variable costs for small, medium and large electric distribution utilities[105].

Table 9.9 Marginal and average variable costs for small, medium and large utilities

Size of utilities	small y = 73,350,000 kWh	medium y = 99,500,000 kWh	large y = 307,350,000 kWh,
Marginal variable cost	0.11	0.10	0.10
Average variable cost	0.12	0.11	0.11

105 The average and marginal total costs have been evaluated at the input prices and load factor of the median company.

Marginal and average variable costs ranges from Swf. 0.09 to Swf. 0.11, while average total cost ranges from Swf. 0.11 to Swf. 0.13. Both are decreasing in output which imply the "L-shaped" average and marginal variable cost curves presented in Figure 9.1.

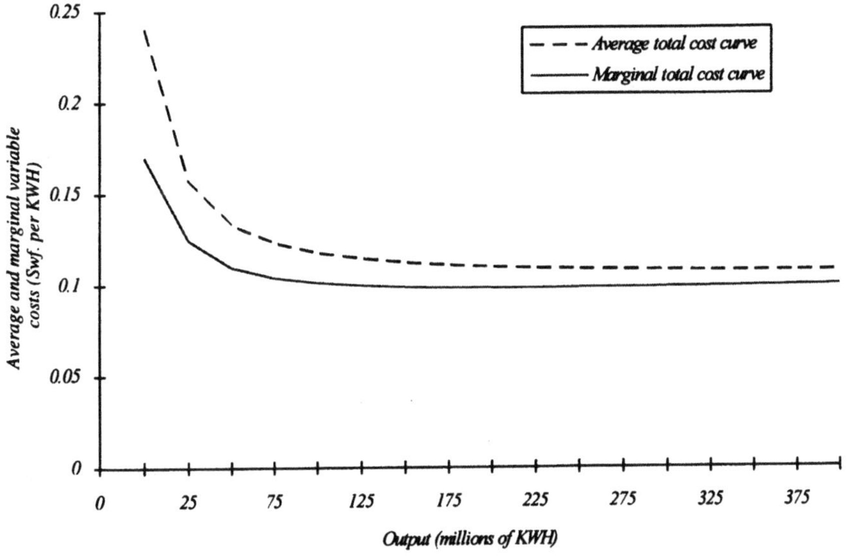

Figure 9.1 Average and marginal variable cost curves

9.7 Elasticities of factor demand and substitution

In this section we presents the short-run own ($SOPE_{jt}$) and cross-price ($SCPE_{jt}$) input elasticities and short-run Allen elasticities of substitution ($SAES_{jt}$) evaluated at the sample median. For the translog variable cost function (9.4) these elasticities are computed using the following formulas:

$$SCPE_{ji} = (\mu_{ji} + S_j S_i) / S_i \qquad \textit{for all } j \neq i$$
$$SOPE_{ii} = (\mu_{ii} + S_i S_i - S_i) / S_i \qquad \textit{for all } i \qquad (9.15)$$

$$SAES_{ji} = (\mu_{ji} + S_j S_i) / S_i S_j \qquad \textit{for all } j \neq i$$
$$SAES_{ii} = (\mu_{ii} + S_i S_i - S_i) / S_i^2 \qquad \textit{for all } i \qquad (9.16)$$

The results reported in Tables 9.10 and 9.11 indicate that all price and substitution elasticities have the correct sign. Moreover, the estimated values of own-price elasticities of factor demand show that labor and electricity demand are inelastic. Further, the positive signs of the Allen elasticity of substitution between electricity and labor indicate that substitution between these two inputs is possible. These results confirm the properties of inputs demand derived from the total cost function which were presented in chapter 8.

Table 9.10 Short-run own and cross-price elasticities of factor demand

	Labor	Electricity
Labor	-0.64	0.09
Electricity	0.64	-0.09

Table 9.11 Short-run Allen elasticities of substitution

	Labor	Electricity
Labor	-4.95	0.73
Electricity		-0.11

9.8 Concluding comments

The purpose of this chapter has been to analyze the cost structure of the Swiss electricity distribution industry employing a short-run cost model. Information on economies of scale and utilization, on average and marginal variable costs, and on demand input responses to changes in input prices are estimated and compared with the results obtained from the total cost function presented in chapter 8.

For this purpose, a translog variable cost function was estimated using panel data for a sample of 39 municipal utilities during the period 1988-1991. The estimation results indicate a non well behaved variable cost function. Thus, all the information on the technology calculated with the estimates of the translog variable cost function have to be interpreted carefully.

This chapter supports the notion that economies of scale and utilization exist in the distribution of electricity, both for small, medium, and large sized utilities. Therefore, these results confirm the results obtained from the estimation of the total cost function.

CHAPTER 10

SUMMARY AND CONCLUSIONS

TO PART II

This chapter presents some conclusions concerning the implementation of producer theory in analyzing the cost structure of the Swiss electricity distribution companies, the most important empirical findings, and several policy implications.

Understandings concerning the implementation of producer theory

The main objective of this part of the book was to analyze the cost structure of the Swiss electricity distribution utilities. The empirical results of this analysis allowed us to assess economies of scale, average cost, the impact of load factor on cost, and demand input responses to changes in input prices.

Relying on the neo-classical duality of production and cost functions, a long- and a short-run translog cost model was used to infer the underlying structure of production. The latter specification provides for the possibility that electric utilities are not in static equilibrium with respect to their capital stock.

The long- and short-run cost functions were estimated using panel data for four years (1988,1989,1990 and 1991) on 39 electricity distribution utilities. The primary data sources were the Swiss Federal Statistical Office's *Wertschöpfungsstatistik* and the Swiss Federal Energy Office's

Finanzstatistik; additional data were taken from the annual publication of the Swiss Cities Association. Data on the number of employees, maximum peak demand and distribution capacity were obtained from a questionnaire sent to the utilities.

By specifying a long- and short-run cost model for the electricity distribution utilities we were faced with two main problems created by limited information available in our data set. The first problem, regarded the possible ways to consider the heterogeneity of output of the electric utilities in cost models. Whereas the second problem, referred to the possible approach of taking into account, in cost models, unobserved characteristics of the service aerea.

To overcome these problems we decided to introduce as an explanatory variable in the long- and short-run cost function specification of the load factor. Moreover, as advocated by Mundlak (1978), we estimated both total and variable cost functions using a fixed-effects model. The inclusion of firm-specific dummy-variables precludes bias in the coefficients of included explanatory variables due to any correlation between unobserved variables that vary by firm but are constant over time, for example output and service area characteristics, and the measured regressors of the cost function.

The choice to analyze the cost structure of the Swiss electricity distribution utilities employing both total and variable cost functions seem to be appealing because each model specification has a unique weakness. The total cost function assumes that electric utilities are in static equilibrium with respect to the capacity. As presented in chapter 7 there are reasons for believing that Swiss electricity distribution utilities have difficulty in adjusting their distributing capacity in response to demand changes. The variable cost function removes that weakness; on the other hand, its results are likely to be somewhat biased because of a multicollinearity problem between the output and the capital stock variable. Further, this econometric problem seems to be at the origin of the counterintuitive sign of the estimated coefficient of capital stock, which indicates a negative marginal product of the capital stock.

The implementation of the production and cost theory to the Swiss electricity distribution sector leads us to the following conclusions:

- the estimation of cost functions using panel data and employing a fixed-effects model represents a valid approach to solve, albeit partially, the problem of output heterogeneity and of unobserved service area characteristics;

- by estimating both long- and short-run cost functions for the electricity distribution utilities one derives a more robust description of the technological characteristics of this sector;

- often, the available data sets do not provide all information to calculate a measure of the capital stock using the perpetual inventory method. Therefore, the applied economist has to utilize proxy variables for the capital goods, which are not completely satisfactory and can give raise to econometric problems.

- to determine the corresponding long-run cost function using the estimation results of a variable cost function and the prices of the capital stock, the variable cost has to be well-behaved and convex in the capital stock. The estimation results of a variable cost function that do not respect these conditions do not allow a precise inference of long-run characteristics of the technology.

Empirical findings

The long- and short-run cost functions for the Swiss electricity distribution utilities were jointly estimated with the parents cost shares equations, using Zellner's (1962) efficient iterative seemingly unreleated regression (SUR) estimation procedure. Moreover, four versions of both long- and short-run cost functions were estimated. Model 1 corresponds to a non-homothetic production technology, while in model 2 the homothetic restriction is imposed. Model 3 represents a Cobb-Douglas cost function, while model 4 restricts the fixed effects so that they are the same for all electric utilities.

The empirical results indicate that the four versions of the long-run cost function are well behaved, i.e., they satisfy the regularity conditions imposed by the cost theory. The estimation results of all four versions of the short-run cost function are, however, less satisfactory in this respect, because they fail to meet the non-increasing condition in the capital stock of a well-behaved variable cost function. Moreover, the results of both cost models suggest that the major advantages of using a fixed-effects model are increased explanatory power of the regressions and preclusion of bias in the coefficients of included explanatory variables due to any correlation between them and unobserved variables.

In the case of the total cost function a Likelihood Ratio (LR) test rejected the hypothesis that the Swiss data on electricity distribution operations reflect a homothetic, a Cobb-Douglas restricted technology model, and that the fixed effects are identical for all electric utilities. Consequently, only the results on the unrestricted technology models with the inclusion of firm-specific dummy variables were employed further in the cost structure analysis of the Swiss electricity distribution utilities. Whereas in the case of the variable cost function, the hypothesis of homotheticity could not be rejected. Therefore, the results of the homothetic version of the variable cost function were used to analyze the technological characteristics of the electric utilities.

Following are summary findings from both variable and total cost functions estimation. First, the empirical results indicate moderate substitutability among inputs and that factor demands are predominantly own-price inelastic. Second, in accord with earlier findings (e.g., Roberts (1986), Nelson and Primeaux (1988) and Salvanes and Tjφota (1994)) this study supports the notion that economies of scale and economies of capacity utilization exist in the distribution of electricity, both for small, medium and large utilities. Table 10.1 presents the estimates of scale economies evaluated with the estimation results from the long- (ES_{TC}) and short-run (ES_{VC}) fixed-effects cost models.

Table 10.1 Summary of economies of scale

Size of utilities	small y = 73,350,000 kWh	medium y = 99,500,000 kWh	large y = 307,350,000 kWh
ES_{TC}	1.21	1.22	1.25
ES_{VC}	1.11	1.12	1.15

The results indicate that the small, medium and large company can simultaneously increase the production of output and total cost will increase less than proportionally. Furthermore, the importance of scale economies increases with size. Third, this study finds that improvement in load factor can significantly reduce total cost. Finally, the estimation results of the variable cost function are inconclusive regarding the over-capitalization hypothesis in this industry because the multicollinearity problem between the output and the capital stock variable precluded a correct test of it.

Policy implications

The estimated cost functions can help to clarify the efficiency of side-by-side competition at all points of a given service territory versus monopolistic provision of electric power. The empirical evidence suggests subadditivity of the cost function over the relevant range of output for the Swiss electricity distribution utilities, a necessary condition for a natural monopoly to exist.[106] Therefore, the findings show that the cost of serving a market of size S over a municipal territory with one utility is lower than the cost of serving the same market with n competitive utilities which install parallel facilities everywhere. This means that the regulator has no incentive to have several electricity distribution utilities when the aggregate quantity of electricity could be distributed more cheaply by a single utility. Thus, side-by-side competition is less cost-efficient than the monopolistic distribution of electric power and deregulation is not likely to yield efficiency gains in the distribution of power. In general, at the distribution level, the companies should continue to operate as public or local franchised monopolies with legally-defined services territories.[107]

However, to additionally motivate electric utilities to increase cost efficiency, more emphasis should be put on promoting franchise competition, which has been analyzed by Demsetz (1968) and by

---

106 For a discussion on possibilities and implications of electric utilities deregulation in US see Joskow and Schmalensee (1983) and Primeaux (1986).

107 This conclusion is made under the maintained assumption of cost efficiency of the Swiss electric distribution utilities. Supposing that behavior of these utilities is not fully cost minimizing, then, it might be possible that the cost advantages from economies of scale under franchised monopoly are lower than those due to competition.

Schmalensee (1979). Instead of assigning a franchise without an auction, this method introduces competition into the markets and is designed to allow electricity utilities to compete for a franchise to serve the market. One possibility would be that the municipalities request bids to operate as local franchised monopolies in the distribution of electricity. A municipality announces that it will accept bids from all electricity distribution companies that are willing and able to distribute electricity in its community. After which the municipality sets certain service standards and then accepts the lowest bid. The winning electric utility is granted permission to operate as local monopoly in its legally defined service territories and is required to respect the service standards defined by the municipality.

This form of franchise competition assumes that the electricity distribution market satisfies the criteria for a contestable market, as developed by Baumol (1982) and Baumol, Panzar, and Willig (1982). This is described as a market with free entry, where if a monopolistic electric utility does not behave as if it were subject to competition, other firms will enter the market to challenge it and attempt to skim the monopoly profits. Under these conditions, the monopolistic utility will be forced to produce and to price efficiently. Of course, a contestable market can exist only in the absence of barriers to entry and exit, in other words, sunk cost should be zero. However, in the electricity distribution industry, due to large initial investment requirements, sunk costs are not zero. Therefore, the market for electricity distribution is not a contestable market and franchise competition cannot be utilized in the form presented above.

Another form of franchise competition that does not present the sunk cost problem is operating franchise competition. This method of promoting competition to serve the electricity market implies that the ownership of capital assets, including transmission lines, installed distributing capacity as well as other infrastructure facilities is separate from actual operations and is held by the municipality. As a result, the municipalities request bids to operate as a local franchised monopoly in the distribution of electricity only for managerial and operating functions, but the municipalities remain responsable for investments decisions.

In addition, our results indicate that an efficient method of decreasing units costs is by increasing the load factor of the electricity distribution system. Policies to spread the peak electricity demand over a longer time period are recommended. For instance, time-of-day pricing, which encourages electricity consumption to be shifted to periods when excess capacity is available, could improve the load factor and thus reduce unit costs. A widespread introduction of this pricing policy could help solve the peak-usage problem facing the Swiss electricity distribution utilities. This strategy is supported by the results on residential electricity demand by time-of-use presented in the first part of the book, which show that electricity time-of-use rates can have the effect of shifting residential electricity consumption from peak to off-peak hours. Moreover, from a commercial point of view we share the old but perhaps actual idea of Parson (1939):

"From a commercial point of view therefore the ...cheapness with which an undertaking can afford to supply electricity depends largely on the magnitude of its load factor" .

The results on economies of scale show that the majority of the electricity distribution utilities operate at an inappropriately low scale. The market territory area of most of these utilities appear too small to produce at optimal scale. Therefore, mergers between two small electricity distribution companies whose market territories are adjacent would increase the geographical coverage and the number of customers of a single electricity distribution system and therefore, improve the scale efficiency of these companies. In general, the main cost savings of a merger between two electricity distribution companies whose market territories are adjacent are likely to be associated with the administrative functions of the company. However, given the diversity of customers and their geographical location, the consolidation of small local companies into regional companies could bring about a reduction in the peak-load pressure on the system and a better use of its capacity. Moreover, there may be opportunities for reduction of redundant facilities and for more efficient utilization of the equipment and operating crews with respect to total electric utiliy size. Furthermore, such mergers could induce an electricity network consolidation and an electricity flow consolidation. The first type of consolidation implies some reduction in the physical network, either by changing the network configuration, or by an abandonment of an existing distributor station. Whereas the second type of consolidation implies concentrating the flows of electricity over the network, i.e., reducing the number of transmission lines. These consolidations could contribute to a further improvement of capacity utilization.

In our sample we have found four constellations of electricity distribution utilities whose market territories are adjacent and for which mergers would be feasible. (Due to the confidentiality of our data, we cannot present the name and descriptive statistics of these eight utilities). Using the estimation results of the long-run cost function presented in chapter 8, we were able to simulate mergers between pairs of these eight electric utilities by comparing the sum of predicted costs for a given pair of electric utilities with the predicted cost of the merged entity whose output is the sum of the output of the two electricity utilities and whose input prices and load factor are the quantity weighted input prices and load factor facing each utility. Table 9.2 presents percentage changes in predicted costs due to the realization of these four hypothetical mergers. Such an analysis is useful in predicting patterns of mergers, to the extent that potential consolidation will be influenced largely by cost considerations.

Table 10.2 Percentage change in predicted total costs due to merger (specific pairs)

Electric utility pair	A/B	C/D	E/F	G/H
% change in predicted cost	10.1	13.8	9.2	15.3

The results reported in Table 10.2 show that the potential cost saving effects of mergers range between 9 and 15 percent.

When evaluating whether a particular electricity distribution utility merger is in the public interest it cannot be forgotten that a merger could be deemed harmful to the public if it increased electric utility market power and allowed electric utilities to set rates greater than costs without losing substantial demand for their commodity.[108] However, in the case of the

108 See Williamson (1968) for a formalized presentation of the welfare effects of mergers.

Swiss electricity distribution utilities of our sample this problem is non-existent, because rates reforms have to be approved by a public rates commission. Therefore, we can conclude from this simple simulation exercise that mergers of pairs of Swiss municipally electricity distribution utilities could be in the interest of the customers living in these communes. Moreover, a process of efficient mergers between small and medium utilities could be induced by grouping the service area of several adjacent municipalities. In this case, instead of announcing several bids for small service territories, a group of municipalities announce only one bid for a large area to be serviced.

Suggestions for further research

The results presented in the first and last part of the book, suggest avenues for further analysis of other elements of the Swiss electricity market, such as the commercial and industrial electricity demand by time-of-use, the cost structure of electric generation and vertically integrated utilities and the existing rates structure. Also, it would be useful to analyze the efficiency of electricity prices. For example, by performing an empirical test for the existence of Ramsey pricing in this industry. This test requires data on the prices, marginal costs, and demand elasticities of residential, commercial and industrial customer classes.

From a larger energy policy perpective, it would be useful, using the empirical results on all elements of the Swiss electricity market and information concerning recent experiences with the reorganization of the electricity industry in United Kingdom and in Norway, to outline a model of a reorganized electricity market for Switzerland.

APPENDIX A

Table A. Electricity demand estimated coefficients of the city specific dummy variables (standard errors in parantheses)

City	*Non-Homothetic AIDS*	*Homothetic AIDS*
F1	-0.319	-0.308
	(0.017)	(0.011)
F2	-0.202	-0.200
	(0.017)	(0.011)
F3	-0.086	-0.104
	(0.018)	(0.019)
F4	-0.088	-0.093
	(0.013)	(0.015)
F5	-0.098	-0.089
	(0.009)	(0.009)
F6	-0.125	-0.121
	(0.007)	(0.007)
F7	-0.062	-0.080
	(0.009)	(0.004)
F8	-0.008	-0.079
	(0.014)	(0.016)
F9	0.017	0.005
	(0.018)	(0.019)
F10	-0.175	-0.174
	(0.006)	(0.007)
F11	-0.027	-0.072
	(0.021)	(0.009)
F12	-0.107	-0.114
	(0.010)	(0.012)
F13	-0.031	-0.039
	(0.027)	(0.030)
F14	-0.126	-0.133
	(0.016)	(0.018)
F15	-0.151	-0.156
	(0.006)	(0.007)
F16	-0.296	-0.291
	(0.016)	(0.015)
F17	-0.202	-0.059
	(0.017)	(0.011)
F18	-0.086	-0.005
	(0.018)	(0.012)
F19	-0.193	-0.176
	(0.015)	(0.016)
F20	-0.018	-0.045
	(0.013)	(0.006)

APPENDIX B

Table B1 Total cost estimated coefficients of the firm specific dummy variables (standard errors in parantheses)

Firm	Model 1	Model 2	Model 3
F1	-0.294	-0.276	-0.474
	(0.030)	(0.030)	(0.027)
F2	-0.401	-0.385	-0.729
	(0.036)	(0.037)	(0.032)
F3	-0.279	-0.270	-0.523
	(0.035)	(0.036)	(0.032)
F4	0.256	0.224	0.557
	(0.057)	(0.056)	(0.034)
F5	-0.046	-0.050	-0.122
	(0.008)	(0.008)	(0.007)
F6	-0.213	-0.203	-0.313
	(0.024)	(0.024)	(0.022)
F7	-0.182	-0.178	-0.265
	(0.011)	(0.011)	(0.009)
F8	-0.184	-0.176	-0.385
	(0.019)	(0.020)	(0.018)
F9	-0.289	-0.280	-0.492
	(0.032)	(0.033)	(0.029)
F10	-0.251	-0.241	-0.352
	(0.018)	(0.019)	(0.018)
F11	0.389	0.337	0.710
	(0.081)	(0.087)	(0.043)
F12	-0.270	-0.259	-0.355
	(0.020)	(0.020)	(0.019)
F13	-0.385	-0.377	-0.479
	(0.022)	(0.022)	(0.020)
F14	-0.250	-0.239	-0.337
	(0.017)	(0.017)	(0.016)
F15	-0.256	-0.241	-0.572
	(0.035)	(0.035)	(0.032)
F16	0.128	0.080	0.291
	(0.047)	(0.046)	(0.029)
F17	-0.230	-0.218	-0.492
	(0.033)	(0.033)	(0.029)

F18	-0.403	-0.394	-0.576
	(0.026)	(0.026)	(0.025)
F19	0.159	0.122	0.264
	(0.031)	(0.030)	(0.021)
F20	-0.276	-0.267	-0.326
	(0.025)	(0.026)	(0.025)
F21	0.044	0.041	-0.048
	(0.009)	(0.009)	(0.008)
F22	-0.003	-0.020	0.011
	(0.012)	(0.011)	(0.009)
F23	-0.186	-0.176	-0.299
	(0.019)	(0.019)	(0.017)
F24	-0.358	-0.347	-0.484
	(0.018)	(0.018)	(0.017)
F25	-0.294	-0.285	-0.467
	(0.028)	(0.028)	(0.026)
F26	-0.239	-0.226	-0.252
	(0.023)	(0.024)	(0.021)
F27	-0.031	-0.041	0.048
	(0.016)	(0.016)	(0.012)
F28	-0.206	-0.194	-0.343
	(0.022)	(0.022)	(0.019)
F29	-0.294	-0.276	-0.586
	(0.041)	(0.041)	(0.036)
F30	-0.173	-0.188	-0.081
	(0.012)	(0.012)	(0.009)
F31	0.206	0.171	0.350
	(0.038)	(0.037)	(0.025)
F32	-0.244	-0.235	-0.386
	(0.019)	(0.020)	(0.018)
F33	-0.184	-0.175	-0.350
	(0.019)	(0.020)	(0.018)
F34	-0.407	-0.399	-0.649
	(0.025)	(0.025)	(0.023)
F35	0.089	0.076	0.129
	(0.014)	(0.014)	(0.011)
F36	-0.258	-0.248	-0.226
	(0.012)	(0.012)	(0.011)
F37	-0.269	-0.254	-0.584
	(0.033)	(0.034)	(0.030)
F38	-0.009	-0.024	0.020
	(0.013)	(0.012)	(0.010)

Table B2. Variable cost estimated coefficients of the firm specific dummy variables (standard errors in parantheses)

Firm	Model 1	Model 2	Model 3
F1	-0.221	-0.209	-0.160
	(0.033)	(0.032)	(0.027)
F2	-0.290	-0.285	-0.293
	(0.037)	(0.037)	(0.030)
F3	-0.232	-0.227	-0.225
	(0.033)	(0.032)	(0.028)
F4	0.121	0.122	0.091
	(0.052)	(0.052)	(0.032)
F5	0.010	0.011	-0.026
	(0.006)	(0.006)	(0.005)
F6	-0.185	-0.183	-0.173
	(0.021)	(0.021)	(0.019)
F7	-0.181	-0.179	-0.167
	(0.009)	(0.009)	(0.008)
F8	-0.095	-0.093	-0.093
	(0.019)	(0.018)	(0.016)
F9	-0.224	-0.219	-0.225
	(0.032)	(0.032)	(0.027)
F10	-0.212	-0.210	-0.210
	(0.019)	(0.019)	(0.016)
F11	0.227	0.232	0.184
	(0.071)	(0.071)	(0.040)
F12	-0.250	-0.249	-0.238
	(0.018)	(0.018)	(0.017)
F13	-0.368	-0.368	-0.361
	(0.020)	(0.020)	(0.018)
F14	-0.211	-0.210	-0.203
	(0.016)	(0.016)	(0.015)
F15	-0.130	-0.124	-0.128
	(0.036)	(0.035)	(0.030)
F16	0.143	0.147	0.122
	(0.042)	(0.042)	(0.025)
F17	-0.197	-0.194	-0.185
	(0.029)	(0.029)	(0.026)
F18	-0.369	-0.369	-0.368
	(0.025)	(0.025)	(0.022)
F19	0.153	0.155	0.136
	(0.026)	(0.026)	(0.018)
F20	-0.222	-0.219	-0.210
	(0.023)	(0.023)	(0.021)

F21	0.097	0.097	-0.109
	(0.008)	(0.008)	(0.006)
F22	-0.044	-0.045	-0.044
	(0.011)	(0.011)	(0.009)
F23	-0.143	-0.141	-0.117
	(0.018)	(0.018)	(0.016)
F24	-0.302	-0.306	-0.296
	(0.020)	(0.020)	(0.019)
F25	-0.232	-0.228	-0.234
	(0.028)	(0.028)	(0.024)
F26	-0.182	-0.181	-0.143
	(0.022)	(0.022)	(0.019)
F27	-0.032	-0.030	-0.032
	(0.015)	(0.015)	(0.010)
F28	-0.180	-0.179	-0.156
	(0.020)	(0.020)	(0.017)
F29	-0.218	-0.212	-0.177
	(0.039)	(0.038)	(0.032)
F30	-0.018	-0.014	-0.058
	(0.011)	(0.011)	(0.007)
F31	0.175	0.177	0.155
	(0.032)	(0.032)	(0.022)
F32	-0.224	-0.222	-0.218
	(0.018)	(0.017)	(0.016)
F33	-0.166	-0.164	-0.157
	(0.017)	(0.017)	(0.015)
F34	-0.339	-0.335	-0.335
	(0.024)	(0.024)	(0.021)
F35	0.072	0.072	0.069
	(0.012)	(0.012)	(0.010)
F36	-0.230	-0.230	-0.225
	(0.011)	(0.011)	(0.010)
F37	-0.137	-0.132	-0.139
	(0.033)	(0.033)	(0.028)
F38	-0.038	-0.038	-0.041
	(0.011)	(0.011)	(0.009)

References

Aigner, S.E. (1984). "The Welfare Econometrics of Peak-Load-Pricing for Electricity." *Journal of Econometrics* 26: 1-15.

Alston, J.M., and Green, R. (1990). "Elasticities in AIDS Models." *American Journal of Agricultural Economics* 72: 442-445.

Alston, J.M., Foster, K.A., and Green, R. (1994). "Estimating Elasticities with Linear Approximate Almost Ideal Demand System: Some Monte Carlo Results." *Review of Economics and Statistics* 76: 351-356.

Anderson, D. and Turvey, R. (1977). *Electricity Economics.* The John Hopkins University Press.

Atkinson, S. E. (1979). "Responsiveness to Time-of-Day Electricity Pricing." *Journal of Econometrics* 9: 79-95.

Balestra, P. and Nerlove, M. (1966). "Pooling Cross-Section and Time Series Data in the Estimation of a Dynamic Model: The Demand for Natural Gas." *Econometrica* 34: 585-612.

Baltagi, B.H. and Griffin, J.M. (1984). "Short and Long-run Effects in Pooled Models." *International Economic Review* 25: 631-645.

Baltagi, B.H. and Raj, B. (1992). "A Survey of Recent Theoretical Developments in the Econometrics of Panel Data." *Empirical Economics* 17: 85-109.

Barnett, A.P. (1969). "Maximum Likelihood Estimation of a Complete System of Demand Equations." *European Economic Review* 1: 25-29.

Barten, A.P. (1969). "Maximum Likelihood Estimation of a Complete System of Demand Equations." *European Economic Review* 1: 25-29.

Barten, A.P. (1993). "Consumer Allocation Models: Choice of Functional Form." *Empirical Economics* 18: 129-158.

Baumol, W.J. (1982)." Contestable Markets: an Uprising in the Theory of Industry Structure." *American Economic Review* 72: 178-183.

Baumol, W.J., Panzar, J.C. and Willig, R.D. (1982). *Contestable Markets and the Theory of Industry Structure*. Harcourt Brace Jovanovich.

Beck, K. and Zweifel, P. (1988). "Warum eine Grenzkostentarifierung für Elektrizität?." *Schweizerische Zeitschrift für Volkswirtschaft und Statistik* 4: 543-558.

Beck, K. and Zweifel, P. (1994). "Optimal Pricing and Provision of Electricity in Markets with Trade." Mimeo, University of Zurich.

Belsey, D., Kuh, E. and Welsch, R. (1980). *Regression Diagnostics: Identifying Influential Data and Sources of Collinearity*. Wiley.

Berndt, E.R. and Morrison, C. (1981). "Capacity Utilization Measures: Underlying

Economic Theory and an Alternative Approach." *American Economic Review* 71: 48-52.

Berndt, E.R. (1991). *The Practice of Econometrics.* Addison Wesley.

Binswanger, H.P. (1974). " A Cost Function Approach to the Measurement of Elasticities of Factor Demand and Elasticities of Substittution." *American Journal of Agricultural Economics* 56: 377-386.

Blackorby, C., Primont, D. and Russel, R.R. (1978). *Duality, Separability and Functional Structure.* American Elsevier.

Blisard, W. and Blaylock, J. (1991). "Construction of True Cost of Food Indexes from Estimated Engel Curves." *American Journal of Agricultural Economics* 73: 775-783.

Blundell, R. (1988). "Consumer Behavior: Theory and Empirical Evidence - A Survey." *Economic Journal* 98: 16-65.

Blundell, R., Pashardes, P. and Weber G. (1993). "What Do We Learn About Consumer Demand Patterns from Micro Data ." *American Economic Review* 83: 570-597.

Braeutigam, R.R. and Daughety A.F. (1983)." On the Estimation of Returns to Scale using Variable Cost Functions." *Economics Letters* 11: 25-31.

Browning, M. (1993). "Estimating Micro Parameters from Macro Data Alone: Some Pessimistic Evidence." *Ricerche Economiche* 47: 253-267.

Bruse, H. and Fuhrmann, G.H. (1981). "Capital Stock of Firms - Calculation on a Microeconomic Basis." *Zeitschrift für Nationalökonomie* 41: 361-383.

Callan, S.J. (1991). " The Sensitivity of Productivity Growth Measures to Alternative Structural and Behavioral Assumptions: An Application to Electric Utilities 1951-1984."*Journal of Business & Economics Statistics* 9: 207-213.

Carlevaro, F. and Spierer, C. (1983)." Dynamic Energy Demand Models With latent Equipment." *European Economic Review* 23: 161-194.

Carlevaro, F., Chaze, J.P., Krishnakumar, J. and Loza, H. (1994). *La consommation des ménages en Suisse.* Bundesamt für Konjukturfragen.

Caves, D. W. and Christensen, L. R. (1980)."Econometric Analysis of Residential Time-of-Use Electricity Pricing Experiments." *Journal of Econometrics* 14: 287-306.

Caves, W.C., Christensen, L.R. and Swanson, J.A.(1981)."Productivity Growth, Scale Economies, and Capacity Utilization in U.S. Railroads, 1955-74." *American Economic Review* 71: 994-1002.

Caves, D. W., Christensen, L.R., Schoech, P.E. and Herriges, J.A. (1984a)."A Comparaison of Different Methodologies in a Case Study of Residential Time-of-Use Electricity Pricing." *Journal of Econometrics* 26: 17-34.

Caves, W.C., Christensen, L.R. and Tretheway, M.W. (1984b). " Economies of Density Versus Economies of Scale:Why trunk and Local Service Airlaine Costs Differ." *Rand Journal of Economics* 15: 471-489.

Caves, D. W., Christensen, L. R. and Herriges, J.A. (1987)."The Neoclassical Model of Consumer Demand with Identically Priced Commodities: an Application to Time-of-Use Electricity Pricing." *Rand Journal of Economics* 18: 564-580.

Caves, D. W. and Christensen, L. R. (1988)."The Importance of Economies of Scale, Capacity Utilization, and Density in Explaining Interindustry Differences in Productivity Growth." *The Logistics and Transportation Review* 24: 3-31.

Chalfant, J.A. (1984). "Comparaison of Alternative Functional Forms with Application to Agricultural Data."*American Journal of Agricultural Economics* 66: 216-220.

Chalfant, J.A. (1987). "A Globally Flexible, Almost Ideal demand System."*Journal of Business and Economic Statistics* 5: 233-242.

Chambers, R.G. (1988). *Applied Production Analysis*. Cambridge University Press.

Christensen, L. R. and Jorgenson, D. (1969)."The Measurement of US Real Capital Input, 1929-1967." *Review Income and Wealth Series* 16: 293-320.

Christensen, L. R., Jorgenson, D. and Lau, L.J. (1971)."Trascendental Logarithmic Production Frontiers." *The Review of Economics and Statistics* 55: 28-45.

Christensen, L. R. and Greene, W.H. (1976)." Economies of Scale in U.S. Electric Power Generation." *Journal of Political Economy* 84: 655-676.

Cowing, T.G. and Smith, V.K. (1978)."The Estimation of a Production Technology: A Survey of Econometric Analyzes of Steam-Electric Generation." *Land Economics* 54: 157-186.

Cowing, T.G. and Holtmann, A.G. (1983)."Multiproduct Short-run Hospital Cost Functions: Empirical evidence and Policy Implications from Cross-Section Data." *Southern Economic Journal* 49: 637-653.

Cornes, R. (1992). *Duality and Modern Economics*. Cambridge University Press.

Deaton, A. and Muellbauer, J. (1980a)."An Almost Ideal Demand System." *American Economic Review* 70: 312-326.

Deaton, A. and Muellbauer, J. (1980b). *Economics and Consumer Behavior.* Cambridge University Press.

Demsetz, H. (1968)." Why Regulate Utilities? " *Journal of Law and Economics* 11: 55-65.

Dennerlein, R. K.-H. (1990). *Energieverbrauch privater Haushalte.* MaroVerlag.

Eakin B. K., McMillen D.P. and Buono M.J. (1990). "Constructing Confidence Intervals using the Bootstrap: An Application to a Multi-Product Cost Function." *Review of Economic and Statistics* 72: 339-344.

Filippini, M. (1991). La struttura dei costi delle ferrovie private svizzere secondo la teoria del duale: implicazioni per una politica delle fusioni. unpublished PhD thesis, University of Zurich.

Filippini, M. and Maggi, R. (1993)." Efficiency and Regulation in the Case of the Swiss Private Railways". *Journal of Regulatory Economics* 5: 199-216.

Filippini, M. (1993)."Swiss Residential Demand for Electricity by Time-of-Use: An Application of the Almost Ideal Demand System." *Working paper 9307, University of Zürich.*

Filippini, M. (1994). "Residential demand for electricity by time-of-use: a share equation approach". *Economia delle fonti di energia e dell'ambiente* 20: 7- 24.

Filippini, M. and Schim Van Der Loeff, S. (1994)." Electricity Demand by Time-of-Use: An Application of Exact Aggregation Theory." *METEOR Research Memorandum, University of Limburg.*

Filippini, M. (1995a)."Swiss Residential Demand for Electricity by Time-of-Use: An Application of the Almost Ideal Demand System." *Energy Journal* 16: 1- 13.

Filippini, M. (1995b)."Residential Demand for Electricity by Time-of-Use." *Resource and Energy Economics* 17: 197-204.

Filippini, M. (1996)."Economies of scale and utilization in the swiss electric power distribution industry." *Applied economics* 28: 543-550.

Foreman-Peck, J. and Waterson M. (1985)." The Comparative Efficiency of Public and Private Enterprise in Britain: Electricity Generation between the World Wars." *Economic Journal* 95: 83-95.

Friedlaender, A.F.and Spady, R. (1981). *Freight Transport Regulation: Equity,Efficiency, and Competition in the Rail and Trucking Industries*. MIT Press.

Friedlaender, A.F.and Shaw-Er Wang Chiang, J. (1983). " Productivity Growth in the Regulated Trucking Industry."*Research in Transportation and Economics* 1: 149-184.

Friedlaender, A.F., Berndt, E.R., Shaw-Er Wang Chiang, J., Showalter, M. and Vellturo, C.A.(1993)." Rail Costs and Capital Adjustements in a Quasi -Regulated Environment." *Journal of Transport Economics and Policy* 45: 131-152.

Fuss, M. and McFadden, D. (1978). *Production Economics: A Dual Approach to Theory and Applications* (Vol.1,2). North-Holland.

Fuss, M., McFadden, D. and Mundlak, Y. (1978). "Functional Forms in Production Theory." In *Production Economics: A Dual Approach to Theory and Applications* (Vol.2), eds. Fuss, M. and McFadden, D.. North-Holland.

Gollop, F. and Roberts, M. (1983)." Environmental Regulations and Productivity Growth: The Case of Fossil-Fueled Electric Power Generation." *Journal of Political Economy* 49: 654-674.

Good, D.H., Nadiri, M.I. and Sickles, R.C. (1992)."The Structure of Production,Technical Change and Efficiency in a Multiproduct Industry: An Application to U.S. Airlaines." Working Paper No. 3939, National Bureau of Economic Research, Cambridge.

Gorman, W.M. (1953). "Community Preferences Fields ." *Econometrica* 21: 63-80.

Gorman, W.M. (1961). "On a Class of Preferences Fields ." *Metroeconomica* 13: 53-65.

Gorman, W.M. (1976). "Tricks with Utility Functions." In *Essays in Economic Analysis*, eds. M.Artis and R.Nobay. Cambridge University Press.

Greene, W. H. (1993). *Econometric Analysis*. MacMillan Publishing Company.

Griffin, R.C., Montgomery, J.M., and Rister M.E. (1987). "Selecting Functional Form in Production Analysis." *Western Journal of Agricultural Economics* 22: 216-227.

Guyomard, H and Vermersch D. (1989). "Derivation of Long-Factor Demands from Short-Run Responses." *Agricultural Economics* 3: 213-230.

Hammond, C.J. (1992). "Privatisation and the Efficiency of Decentralised Electricity Generation: Some Evidence from Inter-War Britain." *The Economic Journal* 102: 538-553.

Hanemann, M. and Morey, E. (1992). "Separabily, Partial Demand Systems, and Consumer's Surplus Measures." *Journal of Environmental Economics and Management* 22: 241-258.

Hanoch, G. (1975)." The Elasticity of Scale and the Shape of Average Costs." *American Economic Review* 65: 956-965.

Hausmann, J.A. (1981). "Exact Consumer Surplus and Deadweight Loss." *American Economic Review* 71: 662-676.

Hawdon, D. (1992). *Energy Demand, Evidence and Expectations*. Surrey University Press.

Heien, D. and Pompelli, G. (1989). "The Demand for Alcoholic Beverages: Economic and Demographic Effects ." *Southern Economic Journal* 55: 759-770.

Heineke, J.M. (1993). "Exact Aggregation of Consumer Demand Systems ." *Ricerche Economiche* 47: 215-232.

Henderson, J.S. (1985). "Cost Estimation for Vertically Integrated Firms: The Case of Electricity." In *Analysing the Impact of Regulatory Change in Public Utilities*, eds: Crew, M.A.. Lexington Books.

Hill, D.H., Ott, D.A., Taylor, L.D. and Walker, J.M. (1983). "Incentive Payments in Time-of-Day Electricity Pricing Experiments: the Arizona Experience." *The Review of Economics and Statistics* 65: 59-65.

Hogan, W.W. (1993)." A Competitive Electricity Market Model." Working Paper, Center for Business& Government, John F. Kennedy School of Government, Cambridge.

Hsiao, C. (1986). *Analysis of Panel Data*. Cambridge University Press.

Huettner, D.A. and Landon, J.H. (1974). "Electric Utilities: Economies and Diseconomies of Scale." Working Paper No. 55, Research Program in Industrial Economics, Case Western Reserve University.

Jehle, R.L. (1987). *Advanced Microeconomic Theory*. Prentice-Hall International.

Johnston, J. (1952). "Statistical Cost Functions in Electricity Supply." *Oxford Economic Papers* 7: 68-105.

Jorgenson, D.W. (1963). "Capital Theory and Investment Behavior." *American Economic Review* 53: 247-259.

Jorgenson, D.W. (1967). "The Theory of Investment Behavior." In *Determinants of Investment Behavior*, National Bureau of economic Research. Columbia University Press.

Jorgenson, D.W. (1974). "The Economic Theory of Replacement and Depreciation." In *Econometric and Economic Theory, Essays in Honour of Jan Tinbergen*, eds.: W., Sellekaerts. MacMillan.

Jorgenson, D.W., Lau,L.J., and Stoker T.M. (1982). "The Transcendental Logarithmic Model of Aggregate Consumer Behavior." In *Advances in Econometrics* (Vol.1), eds.: R.L. Basmann and G.F. Rhodes. JAI Press.

Jorgenson, D.W., Slesnick D.T., and Stoker T.M. (1988). "Two-Stage Budgeting and Exact Aggregation." *Journal of Business & Economic Statistics* 6: 313-325.

Joskov, P.L. and Schmalensee, R. (1983). *An Analysis of Electric Utility Deregulation.* MIT Press.

Judge,G., Griffiths, W., Hill, C., Lutkepohl, H. and Lee,T. (1988). *Introduction to the Theory and Practice of Econometrics.* Wiley.

Kamerschen, D.R. and Thompson, H.G., (1993). " Nuclear and Fossil Fuel Steam Generation of Electricity :Differences and Similarities." *Southern Economic Journal* 48: 687-700.

Koutsoyiannis, A. (1979). *Modern Microeconomics.* MacMillan.

Kmenta, J. and Gilbert, R.F. (1968)."Small Sample Properties of Alternatives Estimators of Seemingly Unrelated Regressions." *Journal of American Statistical Association* 63: 1180-1200.

Kohler, D.F. (1983). "The Bias in Price Elasticity Estimates Under Homothetic Separability: Implications for Analysis of Peak-Load Electricity Pricing." *Journal of Business and Economic Statistics* 1: 202-210.

Kukatilaka, N. (1985). "Tests on the Validity of Static Equilibrium Models." *Journal of Econometrics* 28: 253-263.

Krinsky, I. and Robb A.L. (1986). "On Approximating the Statistical Properties of Elasticities." *Review of Economic and Statistics* 68: 715-719.

LaFrance, J.T. (1991)."When Is Expenditure Exogenous in Separable Demand Models?" *Western Journal of Agricultural Economics* 16: 49-62.

LaFrance, J.T. (1993)." Weak Separability in Applied Wefare Analysis." *American Journal of Agricultural Economics* 75: 770-775.

LaFrance, J.T. and Hanemann, W.M. (1989)."The Dual Structure of Incomplete Demand System." *American Journal of Agricultural Economics* 71: 262-274.

Larsen, R.J. and Marx, M.L. (1986). *An Introduction to Mathematical Statistics and its Applications.* Prentice-Hall.

Lau, L.J. (1976). "A Characterization of Normalized Restricted Profit Function." *Journal of Economic Theory* 12: 131-163.

Lau, L.J. (1982). "A Note on the Fundamental Theorem of Exact Aggregation." *Economic Letters* 9: 119-126.

Maddala , G.S. (1971)." The Likelihood Approach to Pooling Cross-section and Time Series Data." *Econometrica* 39: 939-953.

McKenzie, G.W. (1983). *Measuring Economic Welfare.* Cambridge University Press.

Mdafri, A. and Brorsen B. Wade (1993). "Demand for Red Meat, Poultry, and Fish in Morocco: an Almost Ideal Demand System." *Agricultural Economics* 9: 155-163.

Meier, M. (1993). "Elektrizitätstarifierung: Theorie und Empirie." *Mimeo*, University of Zurich.

Morrison,C. (1985)." Primal and Dual Capacity Utilization: An Application to Productivity Measurement in the U.S. Automobile Industry." *Journal of Business & Economics Statistics* 3: 312-324.

Mountain, D.C. and Lawson, E.L. (1992). "A Disaggregated Non-homothetic Modeling of Responsiveness to Residential Time-of-Use Electricity Rates." *International Economic Review* 33: 181-207.

Muellbauer, J. (1975). "Aggregation, Income Distribution and Consumer Demand." *Review of Economic Studies* 62: 525-543.

Muellbauer, J.(1976). "Community Preferences and the Representative Consumer." Econometrica 44, 979-999.

Mundlak, Y. (1978)." On the Pooling of Times-Series and Cross-Section Data." *Econometrica* 46: 69-85.

Mutzner, J. (1993). "Schweizerische Stromtarife." *Bulletin SEV/VSE* 80: 35-45.

Nelson, R.A. (1985)." Return to Scale from Variable and Total Cost Functions." *Economic Letters* 18: 271-276.

Nelson, R.A. (1989)." On the Measurement of Capacity Utilization."*Journal of Industrial Economics* 37: 273-286.

Nelson, R.A. (1990)."The Effects of Competition on Publicly-Owned Firms, Evidence from the Municipal Electric Industry in the U.S." *International Journal of Industrial Organisation* 8: 37-51.

Nelson, R.A. and Primeaux, W.J., Jr. (1988)." The Effects of Competition on Transmission and Distribution Costs in the Municipal Electric Industry." *Land Economics* 64: 338-346.

Nerlove, M. (1963). "Return to Scale in Electricity Supply." In *Measurement in Economic Studies in Honor of Yehuda Grunfeld*, eds.: C.F., Christ. Stanford University Press.

Nordin, J. (1947). "Note on a Light Plant's Cost Curves." *Econometrica* 15: 321-335.

Nordin, J. A. (1976). "A Proposed Modification of Taylor's Demand Analysis: Comment." *Bell Journal of Economics* 7: 719-721.

Norsworthy, J.R. and Jang, S.L. (1992). *Empirical Measurement and Analysis of Productivity and Technological Change*. North Holland.

Oum, T.H. and Tretheway, M.W. (1989)."Hedonic versus General Specification of the Translog Cost Function." *The Logistic and Transportation Review* 25: 3-21.

Oum, T.H., Tretheway, M.W. and Zhang, Y. (1991)."A Note on Capacity Utilization and Measurement of Scale Economies." *Journal of Business & Economics Statistics* 9: 119-123.

Panzar, J.C. and Willig, R.D. (1977)."Economies of Scale in Multi-Output Production." *Quality Journal of Economic* 91: 481-493.

Parks, R. W. and Weitzel, D. (1984)." Measuring the Consumer Welfare Effects of Time-Differentiated Electricity Prices." *Journal of Econometrics* 26 , 35-64.

Parsons, R.H. (1939). *Early Days in the Power Station Industry*. CUP for Babcock and Wilcox.

Phlips, L. (1983). *Applied Consumption Analysis*. North-Holland.

Pollak, R.A. and Wales, T.J. (1992). *Demand System Specification and Estimation*. Oxford University Press.

Primeaux, W.J., Jr. (1986). *Direct Electric Utility Competition*. Praeger.

Ray, R. (1980). "Analysis of a Time Series of Household Expenditures Surveys for India." *The Review of Economics and Statistics* 62: 595-602.

Roberts, M.J. (1986). " Economies of Density and size in the Transmission and Distribution of Electric Power." *Land Economics* 62: 378-387.

Rossi, N. (1988)."Budget Share Demographic Translation and Aggregate Almost Ideal Demand System." *European Economic Review* 31: 1301-1318.

Rushdi, A.A. (1991)."Economies of Scale and Factor Substitution in Electricity Supply Industry." *Energy Economics* 15: 219-229.

Salvanes, K.G. and Tjϕota, S. (1994)." Productivity Differences in Multiple Output Industries: An Empirical Application to Electricity Distribution." *The Journal of Productivity Analysis* 5: 23-43.

Samuelson, P. (1974). "Complementary: An Essay on the 40th Anniversary of the Hicks-Allen Revolution in Demand Theory ." *Journal of economic Literature* 64: 1255-1289.

Schmalnsee, R. (1979). *The Control of Natural Monopoly*. Lexington Books.

Shephard, R.W. (1953). *Cost and Production Functions*. Princeton University Press.

Shephard, R.W. (1970). *Theory of Cost and Production Functions*. Princeton University Press.

Solow, R.M. (1957)." Technical Change and the Aggregate Production Function." *The Review of Economics and Statistic* 39: 312-320.

Spierer, Ch. (1988). Modélisation économétrique et perspectives à la long terme de la demande d'énergie en Suisse, Expertengruppe Energieszenarien, Schriftenreihe Nr. 17, Bern.

Swiss Statistical Office (1992). *Verbrauchserhebung 1990*. Bern.

Swiss Energy Office (1994). *Schweizerische Elektrizitätsstatistik 1993*. Bern.

Swiss Statistical Office (various years). Wertschöpfungsstatistik, Bern.

Swiss Energy Office (various years). Finanzstatistik, Bern.

Swiss Cities Association (various years). Statistik der Schweizer Städte, Zürich.

Stern, N. (1986). "On the specification of Labor Supply Functions." In *Unemployment,Search and Labour Supply* eds. Blundell, R. and Walker, I.. Cambridge University Press.

Stewart, J. (1979)." Plant Size, Plant Factor, and the Shape of the Average Cost Function in Electric Power Generation: A Non-homogeneous Capital Approach." *Bell Journal of Economics* 10: 549-565.

Strotz, R.H. (1959). "The Utility Tree - a Correction and Further Appraisal ." *Econometrica* 25: 269-280.

Stoker, T.M. (1993). "Empirical Approaches to the Problem of Aggregation Over Individuals." *Journal of Economic Literature* 31: 1827-1874.

Theil, H. (1967). *Economics and Information Theory*. North-Holland.

Taylor, L. D. (1975)." The Demand for Electricity: A Survey." *Bell Journal of Economics* 6: 74-110.

Taylor, T. N. and Schwarz, P. M. (1990)." The Long-Run Effects of a Time-of-Use Demand Charge." *Rand Journal of Economics* 21: 431-445.

Thomas, R.L. (1987). *Applied Demand Analysis*. Longman.

Thompson, H.G., Jr. and Wolf, L.(1993). "Regional Differences in Nuclear and Fossil-Fuel Generation of Electricity." *Land Economics* 69: 234-248.

Tobin, J.(1950)." A Statistical Demand Function for Food in the USA." *Journal of the Royal Statistical Society, Series A* 113: 113-141.

Wallace,T.D. and Hussain, A. (1969). "The Use of Error Components Models in Combining Cross-Section and Time-Series Data." *Econometrica* 48: 55-72.

Varian, H.R. (1992). *Microeconomic Analysis*. Norton.

Weiss, L.W. (1975). " Antitrust in the Electric Power Industry." In A. Phillips, *Promoting Competition in Regulated Markets*, Washington, DC: Brooking Institution.

Wellman, K.F. (1992). "The US Retail Demand for Fish Products: an Application of the Almost Ideal Demand System." *Applied Economics* 58: 18-36.

Williamson, O.E. (1968). "Economies as a Antitrust Defense: The Welfare Trade-Off." *The American Economic Review* 24: 445-457.

Zellner, A. (1962)."An Efficient Method of Estimating Seemingly Unrelated Regressions and Test for Aggregation Bias." *Journal of the American Statistical Association* 58: 348-368.

GPSR Compliance

The European Union's (EU) General Product Safety Regulation (GPSR) is a set of rules that requires consumer products to be safe and our obligations to ensure this.

If you have any concerns about our products, you can contact us on ProductSafety@springernature.com

In case Publisher is established outside the EU, the EU authorized representative is:

Springer Nature Customer Service Center GmbH
Europaplatz 3
69115 Heidelberg, Germany

Batch number: 10725979

Printed by Printforce, the Netherlands